Falling from Grace
in Texas

*a literary response to the
demise of paradise*

Falling from Grace in Texas

a literary response to the demise of paradise

Edited by

Rick Bass
and
Paul Christensen

Wings Press
San Antonio, Texas
2004

Falling from Grace in Texas:
A literary response to the demise of paradise © 2004
by Wings Press
for Rick Bass and Paul Christensen.
Cover photo, "Smog in the Desert," © 2000 by Bryce Milligan.
All rights to individual works contained in this book
revert to the contributing writers upon publication.

First Wings Press Edition
ISBN: 0-930324-57-9

Wings Press
627 E. Guenther
San Antonio, Texas 78210
Phone/fax: (210) 271-7805
On-line catalogue and ordering:
www.wingspress.com

This publication was made possible in part by a grant
from the Texas Commission on the Arts and the Texas Writers' League
and with the assistance of StoneMetal Press and Gallery.

Library of Congress Cataloging-in-Publication data:

Falling from Grace in Texas:
A literary response to the demise of paradise / edited by
Rick Bass, 1958-
Paul Christensen, 1943-
           p.   cm.
      ISBN: 0-930324-57-9 (paperback)
(non acidic paper)
1. Title. 2. Texas--Literary collections. 3. Natural
resources--Texas--Literary collections. 4. Environmental
policy--Texas--Literary collections.
PS558.T4 F39   2004
808

Contents

Part III: Natural Ways

Epigraph: Elizabeth Crook

Introduction

Paul Christensen

Pioneers **always** behave paradoxically, and Texas' first settlers were no different. Once encamped on virgin soil, their lives depended upon a thorny regime of scrimping and saving the things that sustained their lives – tallow for candles, string for binding up hay, the last few beans in the jar against another winter storm and an ice-bound prairie. What they learned about thrift and putting by gave them their distinctive character as frontier people – hardnosed, practical, realistic about most things. Most things. But with a land of forest and plains unexplored and lying in all directions unused, a certain zeal to waste the infinite plenty overtook many of them. The sod of ten thousand years was broken without a thought of coming dust storms; fences went up in the path of migrating herds; rivers were dammed when a little water would do. Indian wars broke out over even the hint of another raid on a settlement or a lonely ranch. Laying waste to the great expanse of raw nature was as fundamental to pioneer ways as was melting all the candle stubs to fashion a new one.

The new country to such invaders is a wonderland of sorts, but also an ominous and unpredictable habitat, whose dark forests and deserts seem inextricably tied to a sense of demons and monsters, the anger of God, and a curse upon the unwary and the sinful. Down come the trees, and in America, the downing of the great pine belt that reached from Maine to Mexico is one of those epic sagas of destruction that few poets have written about. The Victorian house with its gables and expansive porches, is all that remains of the old wilderness hard wood and pine stands, once the home of black bear and wolf, the legions of wild cats and snakes and all the other citizens of once timeless green worlds.

In the Great American Desert, the old name given to the vast rolling plains and prairies stretching across what we now call the Southwest, one of the wonders of the world existed for a small population of hunters. Thirty million bison roamed a wide grass corridor stretching from western Canada to the short grass perimeters of the lower Pecos River – with spring grass for calving, and broken country up around the Cross Timbers for the winter play and horn-clashing of the young bulls, and the milder and more accessible tall grass country of Kansas for the old bulls to retire to. The bison grounds stretched 2400 miles long and 1200 miles wide, but only 200,000 native people lived and feasted on bison and prong-horn herds – a ratio of humans to nurturing wildlife rarely equaled in our species history. The so-called "Buffalo Culture" of the plains dates from about 1200 A.D., when bison reentered the plains from Mexico after a long period of drought, and hunters established a nomadic civilization invis-ible to outsiders but to the initiated a realm of complicated religion of bison gods and lore, and a cultural encyclopedia of rituals honoring nature in all its forms.

This civilization remains a mystery to most of us. It was a culture of play and games, of tale telling and leisure – so much leisure, in fact, that our myths of Arcadia should be based on Buffalo Culture more than on any Greek model. Arts and crafts abounded, and the young were trained to be spe-cialists in their various trades – from merchants who traveled long distances with pelts and bone jewelry and tools to the old Mexican capitol at Tenochtitlan, to warriors, healers, scouts, and explorers among the men; and among the women, the skills to run a village, to prepare the skins of bisons for the making of tipis, coats and other garments; for the jerking of the beef for the lean winters; the arts of cooking, gathering of herbs and wild food; for healing with medicinal plants; for weaving, basketry, and the rearing of the children. The elders presided over village affairs, and decided when to hunt, or migrate, when to stay put. They were good weathermen and knew how to consult the landscape for the telling of the seasons and the principal events of the circadian calendar. Some were entrusted with the making and protection of temple sights, and for arranging the pictoliths and cave paintings that celebrated their hunts, and the feast days of their year.

As Thomas Mails advised in his study of Plains culture, *The Mystic Warriors of the Plains* (1972), "Those who would know the Indian must study the buffalo intensely too. A broad knowledge of the bison's make-up, habits, and religious connotations is essential to every Plains Indian student."* Consider the buffalo cults involving men and women dressed in the skins and heads of these animals who performed the "Buffalo Dance" and joyfully merged animal and human identity. Shamans wore "buffalo hats" as totemic instruments of magic and healing power. The White Buffalo Woman, an avatar of the *Wakan Tanka* or holy spirit, brought the gift of the Pipe Bundle to the Oglala Sioux, giving the people not only a sacred home in the Plains but the implement of vision for guiding them. The Sun Dance, performed by many Plains nations, featured not only the donning of the Buffalo Hat, but the ritualized reenactment of the impregnation of human beings with wild spirit by the bull god, Erect Horns. Such stories of dancing with gods, gods making love to mortals, transformations of men and women into buffalo spirits rival those of Greek and Egyptian mythology, and are as rich in folklore and vision as any belonging to the European visionary tradition.

Once the Spanish brought horses back into the New World, the ancient traditions of hunting on foot with dogs and travois came to an abrupt halt; after that, native hunters rode bareback ponies in a suddenly uneven, ultimately unfair chase. Fifteen million bison were removed from the gene pool before the white settlers showed up and began thinning and then decimating the remaining herds by the 1880s. By 1900, only about a thousand wild bison remained in America, prompting Teddy Roosevelt to corral a small herd in front of the Smithsonian Institution for Americans to look at, and for the minting of the "Buffalo" nickel in 1913 to revere what

---

* Doubleday & Co., p. 10; see also Mails, *Dog Soldiers, Bear Men and Buffalo Women: A Study of the Societies and Cults of the Plains Indians*. Routledge, 1973.

made America wild and powerful in its innocence. Thirteen years before the Census Bureau proclaimed the closing of the American Frontier, a fact that moved slowly like a dagger through the American unconscious, as if all the raw spirit and strength of a continent and a new people had ebbed away under the squalor of cities.

The exploitation of the land has been relentless, from the predations of the oil and gas industry, to the chemical plants that have been polluting the coastal waters, to the utility plants that have darkened the skies burning cheap coal, natural gas, and lignite, to the weapons factories and nuclear bases, the dude ranches, the mining operations, the airports and the sprawling cities gobbling up the hinterlands in some feud with the openness stretching out to nothing. The car has nearly killed off the ozone canopy and its needs are so great, not even the federal treasury can keep up with the expense of maintaining our asphalt ribbons or of extending them to the last remaining grassy places.

Even so, the memory of the wild plains established itself as a dream in the unconscious of most southwesterners. The "wide open places" haunt us in our blood and is an archetype of what we mean by ultimate freedom and independence. Our treks to Big Bend in the summer and to the many handsome natural parks established in the last century are part of some pilgrimage we must make to the Texas past, that immemorial place that rises in our vision when we feel hemmed in, driven to distraction, depressed by traffic and deadlines and all the tensions and conflicts that arise from life in the city. What European Romantics found in their treks to the Alps, Texans find in their first glimpse of the Staked Plains west of Austin, that rolling ground that spreads itself out to the horizon and silences us with awe. Somewhere out there lies an answer to our longings, a fulfillment of dreams we have had and cannot understand. Something in us is moved by that absence of ourselves, that negation of identity and meaning. Like some Buddhist revelation of the triviality of our meager lives, we look into the face of infinity and see a god of some kind, a powerful natural spirit hovering over us, indifferent to our fates, perhaps, but there as a figure of some unknowable will renewing a lost sense of the spiritual father.

To mourn our abuses of the land, and the disappearing of an old American Eden, tells us we have wandered from some covenant by which people of the past governed themselves and preserved a balance between individual desire and the care of others. We long for the return of such a spirit to reawaken a real sense of bonds to pull us back from the abyss of waste and indulgence, of greedy acquisition and the blind pursuit of riches. Our love of power and money is without purpose, a kind of addictive behavior in which there is never an over dose to warn us of our failing judgment or moral blindness. Out there in the gray ripples of the Plains is that god hiding among the clouds and the struggling grama grasses, and we are too ashamed to stop the car and look for him or it or her.

Instead, writers have come forward since the 1950s to explore just this loneliness we suffer from, and to probe with fine needles into the flesh of our guilt. The Vietnam War precipitated a greater outpouring of such literature of atonement after observing the ravaged jungles scorched by Agent Orange and "carpet bombing," and the mess we left behind us in our rage to invade what seemed like a sudden resurrection of our own Plains nations. That devastation of a nation far away brought home to us the kind of people we had become, a warrior nation bent upon forcing others to our will and into surrendering whatever natural wealth they possessed.

Writers who awaken the conscience are not always welcome in our midst. But we neglect them at our peril, for these are the modern witch doctors who can read our minds and who speak for what we repress in ourselves. And our griefs are written in traffic fatalities, drug abuse, violent crime, our love of violence on TV and in the movies; our passions erupt from our loneliness and shock us when we read about high school shootings, snipers gunning down strangers, our need to own guns and assault rifles, our craving for SUVs and Hummers to ward off the hostilities we expect from other drivers. Writers have been telling us that we are troubled and confused, and have told us our healing can only come through our restraint and our repair of the damage we have wrought on the earth we depend upon. That healing message is written across the breadth of ecological literature, which has told us nature is alive and sensitive and wounded by our rapacity.

In book after book about the earth out of balance, we learn that nature is our own souls, our hidden self – and to ignore it, to starve it, to poison it is eventually to kill ourselves. That other half of human identity was demonstrated to us in the Buffalo Dance, that we are only superficially different from the bull and deer, the hawk and the eagle; we are each of them in part, and because they live in our troubled emotions, they are half of us as well. The hawk on the telephone wire is a figure partly of my own making; he broods over the farmer's weed-shot pasture, over the rusty tractor and bunged roof of the hay shed. He inherits our misuses and stands vigil in the ruins we have made of our habitat.

When Rick Bass put out the call to writers associated one way or another with the Texas Institute of Letters, asking them to contribute to an anthology of protest over the abuses of the Texas environment by developers, by legislators catering to the profiteers of agriculture and petroleum, he intended the book as a gift to each newly elected legislator in the State House at Austin in 2002. Responses came in but not in the torrent he had expected; rather, a trickle of discourse began and built, and sustained itself through 2002 into the present. Word spread, and writers supplied him with pieces that were recently published or had languished in a desk drawer for lack of an audience. The process of gathering, editing, and sorting out the pieces into a coherent whole went on, but something was missing – at least at the start. A honed sense of anger and outrage that would warn the state's Congressmen to watch out, that a tide of collective protest was rising against their business as usual.

Instead of anger, there was remorse and longing, and a desire to name the things of nature, to take inventory and assess what remained of a lost world. Some of the writers included here, Michael Adams, Tracy Daugherty, are political activists, to be sure; others are poets, like Pattiann Rogers, who specialize in an exquisite attention to details, and who evoke the complexities of our dream of nature through what they know about the hidden world of butterflies and field mice, coyotes on their nightly hunts, and what she calls "the family" of connected things, ourselves included. Others simply want to remember another day, like Charlotte Whaley, who came to Dallas back in 1932, and found a plain little town eager to grow. Seventy years later, she lives in a vast metropolis suffering from all the usual ailments: bad air, traffic congestion, real estate booms, all of which require some greater hand to moderate such growth and clean up the "soots, sulfur and carbon dioxides, nitrogen oxides, and mercury" of unregulated commerce.

Olive Hershey knows of a certain place in Wharton County still fertile and in balance, where an old salt dome will shortly become a dump for "petroleum waste." The theme of her piece and of Stephen Harrigan's tribute to the coastal beaches of South Padre Island is of a fallen paradise, repeated over and over across the state. Places that one cherished in youth, or remembered as a haunt of grandparents, a pine thicket or a swimming hole, pasture land or the undammed Brazos, all of them are integers in a moral calculation of the losses we have suffered for lack of a vision that would moderate and seek compromise between commerce and spiritual needs, between an eagerness for material gain and an obligation to protect and care for the earth that feeds us.

The result is perhaps not what Rick Bass expected or even wanted, but something else, a more problematic and intimate response to the question of what is there to protect. The answer comes back in many ways, but the tacit theme underlying all of these pieces is that we do not own nature or the ground we live on, we borrow from its reserves and share some obligation to give back part of what we take. Such an answer tests our notions of individuality and freedom; invoked here are the children who will inherit the state of nature after our own uses, and we must honor them with better guardianship. The dead also preside in the thoughts of the writers, who bear witness to a different Texas than the one we now live in. They hold the memory of some earlier relation, already troubled by the forces let loose in our own time, but still possessing something of its original character. That elegiac mood can be felt throughout this book. Still, others ask the question, who are we as a people if we cannot unite behind a noble cause? What is the meaning of society and of being countrymen and women if we have nothing in common but our desire for wealth and security?

Perhaps what this book is mostly about is the vision that we have come to the end of one form of democracy and found it lacking in cohesion or purpose. We have been liberated from all the forms of servitude from which we Americans escaped in our history of migration, but we have not yet formed a bond equal to those who danced with their gods on this same landscape. We have stood by, passively, as the ground was devoured and we said nothing. Now that it is altered almost beyond repair or recognition, we feel remorse and shame, or at least a vague sense of unfulfilled obligation. Our sprawling suburbs, our poisoned cities, our flimsy efforts at reform and environmental protection leave us naked before this accusing landscape.

Rick Bass kindly invited me to join him in editing this book after reading my memoir of Texas literary life, *West of the American Dream : An Encounter with Texas*. Perhaps he felt I was a kindred soul in the struggle to change a profound and errant philosophy in this region, since I challenged the writers to learn their own local nature and write more boldly about the relation between self and land. I gladly joined him, of course, and did what I could with the manuscripts he sent me – adding a story of my own, and one or two other pieces, typing the final version and sending it off to Bryce Milligan, who had early on expressed an interest in seeing it. The germ of the idea, the inspiration to put such a book before the public are Rick's. He seemed to know it was time for our concerns to be voiced in one place, and he was right.

Falling from Grace in Texas

a literary response to the demise of paradise

Part I:
Waterways

Earth, air and water; plants, animals and people –
it takes all these to make a living place. Past lives
have made us what we are today. Our present lives
are creating the future. It's up to us to do our best
to make wherever we live here and now the best
possible living place for ourselves and all our fellow
creatures.

> – Charlotte Baker Montgomery
> from the Preface to *The Trail North*

Pete A.Y. Gunter

Reflections from the Grass Roots Up

My people came to Texas in the 1840s from Georgia. They settled at Sivill's Bend, on the Red River northwest of Gainesville. It was a dangerous place, surrounded on three sides by Indian Territory, subject to Kiowa and Comanche raids. The land is "breaky": wooded, hilly, cut by spring-fed creeks. They liked it better than the upland prairie.

By the time I appeared, the Gunters had lost their land. But from the age of five I was taken back to it by my father, who wanted me to see what it was like. And to know what the people there were like.

It was a great place for a kid: a wide sandy river, deep wooded bottomlands, high hills with carved limestone cliffs. At night the owls hooted up and down the river. We "ran the trot-lines:" waded long strands of fishing cord set with baited hooks, then sat on the sandbars watching the stars wheel past and talking, as fish jumped on the river, landing, sometimes, with a splash as hard as a rifle shot.

It was idyllic, that wonderful mixture of easy freedom and natural beauty. It surely did breed a sense of roots, of inheritance, of belonging there, on the endless river. But there was something lacking, something not quite complete, even if the owls and an occasional coyote serenaded the dark woods.

It was not the same river my great grandfather knew, or my grandfather. Dad's stories of wolf hunts, or hundred people catfish fries, or raccoons infesting the wild plum thickets or the persimmon groves couldn't hide his, or my, awareness of the game that was no longer there or the deep woods that were being remorselessly cut back, or the mute testimony of springs that no longer flowed.

With hindsight I can see now what had happened. Consider, for example, the original prairies – as described by a great uncle, John Childers:

> I'd already begun to ride ponies when my dad decided to move back to Wood county when I wasn't but four. I still don't recollect much about that, either, but I do know that he put up the first one hoss gin ever in Wood county that year. The next year, he'd sold out his gin at a good profit and moved back to Sivill's Bend, taking four wagonloads of timber to build a house with. Now there's a thing about that country that can't be seen any where today, and that's the grass they had then. It was so high, that when the lumber was unloaded, the men'd have to hunt for it to find it. Actually, it was so high that in later years when I rode man size hosses, the grass'd turn the rowels on my boots, and I'd be topping a hoss about 14 hands high. It was from knee to hip high on an ordinary man, and when we'd be in there, running the mowing machine and putting up hay for the winter, we'd never see the blade. The only way we knowed where the blade was, was by watching the grass right even to the right of where we had the machine. It'd rise up, then lean to one side. It was so thick that it didn't even fall when it was out but leaned on other grass. It only fell when we came along and stacked it for drying. You've seen the big haystacks they have on farms? Well, that's what I mean. Another thing, too, and that's when cattle laid down. You just had to almost stumble over them to find them in the grass.

What John Childers was saying went a lot farther than he realized. Most ranchland grass today ends up cropped ankle short, or lower. Even should some enlightened rancher give his grass a break, there's' not a chance that anyone in Cooke County, Texas, is going to see rolling waves of grass to the horizon, hip high on a grown man on horseback. Not to speak of grass so thick that it will lean on adjacent grass when it is cut.

But when the grass is grazed down and depleted, other things happen. The root networks (once four, five or six feet deep) begin to shallow. Their capacity to hold the soil and the all-important soil moisture is lost. Floods intensify, springs begin to fail, creeks once healthy begin to dry out in the summer heat. Captain R.B. Marcy, on an expedition through North Texas in 1854, found Elm Creek just west of Gainesville to be forty feet wide and two feet deep, full of clear spring-fed water. This was well into June. At that spot, at that time of year, now, you will find only a trickle. Unless there is muddy water from a recent rain.

A great aunt, Lillian Gunter, saw it clearly. In a handwritten essay ("Notes on Cooke County") she wrote, around 1918:

> Sivills Bend is the largest end in the Red River for many miles. It includes a space of ground about twelve by ten miles bounded on the West, North and East by the Red River and on the south by Bear Head Creek, although people in other parts of the county sometimes designate all of the county north of Fish Creek at Sivills Bend. This river bend varies in elevation from 660 feet at high water level on the Red River to nearly 1000 feet at the top of its two high central mounds. Some of the richest land in Texas is to be found in the broad bottomlands that skirt the river from the White Bluffs on the West to the mouth of Bear Head Creek on the East. The rich red lands of these bottoms were not so long ago covered with a prodigious growth of timber, including hickory, pecan, walnut, cedar, ash, mulberry, hackberry, bay, plum, locust, elm, sycamore and some fifteen to twenty varieties of oak, with large cottonwoods, willows and (beach plums?) fringing the river, and persimmons, (chittin?) wood, dogwood and black and red haws fringing the grassy glades that now and then intervened. These forests were draped with festoons of Virginia creepers, wild grapes, mustangs, possum (haws?) ripening from July to October and with briar and sarsaparilla vines which became a flame of red berries in

the fall. The low places in the glades were carpeted with dewberry and blackberry vines. Probably no place in the temperate climate so abounded in wild food as these river bottoms. Consequently they were a sportsman's paradise.

Aunt Lillian was bragging. (After all, it was, and is, a regional tradition.) But she knew what a really magnificent stand of Southern bottomland forest was. And she also knew what was happening to it. Others of her notes spell out the decline of that land in detail.

I will spare the reader a detailed account of those notes. What they boil down to was the damage done by floods (1892, 1908, 1915), which "poisoned" the once fertile farmland with weeds, and sands, and red clay, making it hard to farm and, in some places, impossible. She blamed it on overgrazing and the farming of marginal land upstream. By most accounts she was right.

Today in that valley, you would not find fifteen to twenty species of oaks. (I count five.) Nor will you find hickory, walnut, or mulberry. If there are Virginia creeper or sarsaparilla vines there, they are well hidden. Nor, of course, will you find buffalo, or antelope, or wolves, or badgers, or porcupines, or flocks of ducks so thick they blacken the skies after a gunshot breaks the silence. After the last big flood there, six years ago, landowners had to use bulldozers and backhoes to get down through deposited sand to anything that could be farmed.

It was said of an old Texas politician that his speeches were like a longhorn steer: two points a long ways apart with a lot of bull in between. I don't know what is in between (a too brief sketch, no doubt, of an extended environmental decline). There are, however, two important points.

The first is that you do not have to be from Colorado, or Vermont, or the Pacific Northwest, to be an environmentalist. You can get it from the grass roots up: From familiarity with prairie, ridge rock, and bottomland forest. You can get it from your family.

The second is that many Texans have done so. You will
see it – if you look – in the flowering of Big Thicket
Associations, Hill Country Conservancies, San Jacinto River
Societies. You will see it in the donation of rural lands to the
Nature Conservancy, in the creation of creek corridors/hiking
paths by Texas towns, in Lady Bird Johnson's Wildflower
Center, in environmental easements donated by landowners,
along creeks and property lines. And you will see it in changing attitudes towards clearcutting by Texas lumber companies.

Texas politicians have been slow to catch onto all this.
But they will. It comes from the grass roots up and is irre-
sistible.

Rosemary Catacalos

Homesteaders

for the Edwards Aquifer

They came for the water,
came to its sleeping place
here in the bed of an old sea,
the dream of the water.
They sank hand and tool into
soil where the bubble of springs
gave off hope, fresh and long,
the song of the water.
Babies and crops ripened
where they settled,
where they married their sweat
in the ancient wedding,
the blessing of the water.
They made houses of limestone
and adobe, locked together blocks
descended from shells and coral,
houses of the bones of the water,
shelter of the water.
And they swallowed the life
of the lime in the water,
sucked its mineral up
into their own bones
which grew strong as the water,
the gift of the water.
All along the counties they lay,
mouth to mouth with the water,
fattened in the smile of the water,
the light of the water,

water flushed pure through the
spine and ribs of the birth of life,
the old ocean,
the stone,
the home of the water.

Stephen Harrigan

The Seam of Water and Land

It is a sparkling summer day on Padre Island. For once the
Gulf is blue all the way into shore, and the petals of beach
evening primrose – furling now in the intense midmorning light
– are a vibrant yellow. Fifty yards off-shore a black skimmer
flies low over the water, knifing through the surface with its
glossy red beak.

There is color and brightness everywhere I look, but for
some reason the Padre Island that haunts my imagination is a
monochromatic landscape, a wilderness of damp sand and
rolling gray water. Padre is a narrow barrier island that runs
the length of the lower Texas coast for 113 miles. I first saw the
island at the age of 10, on one of its more moody and desolate
days. It was winter, the sky overcast, the water cloudy with
churned-up sand, the beach deserted except for an occasional
great blue heron facing the ocean as the wind ruffed its luster-
less plumage. I remember the intimidating press of that wind
against my own body and the monotonous roar of the waves
drowning out every sound except the piercing cries of the
gulls. It was an overpowering place frightening in its solitude
and gray solemnity, and perhaps that's why it laid such a last-
ing claim on me.

In those days the human presence on Padre Island was
rather decrepit and forsaken. It was a place of ancient bait
shops, rickety salt-stained piers, and the corroded chassis of
cars that had foundered in deep sand and never been recov-
ered. There were no beach-side condominiums with board-
walks and "tar-removal stations" and parking lots full of late-
model Suburbans with Mexican license plates. If you ordered
fish in the restaurant, it came fried – not blackened or *en
papillote* or mesquite-grilled. Nobody surfed then. Sailboards
had not been invented. The horizon was not studded with off-

shore oil rigs, and there were no dead sea turtles washing onto the sand with plastic lodged in their intestines.

But Padre Island is still where I would go if I were told I had only a few months to live and I wanted to find some location in which to come to a final accounting of myself. There are other places I can think of that might be more inspiring: an overgrown burial dune I once visited in Maui, from whose summit I could see humpback whales rearing out of the ocean; an African plain with the steely slopes of Mount Kenya rising in the background like a massive thundercloud; a glacier in Alaska gleaming with elusive blue light.

I was an adult when I first saw those places. But Padre Island got to me early and gave me a mental template to refer to whenever I have cause to ponder the raw mystery of nature. I remember the island from my childhood as more of a punishing place than a nurturing one, a pitiless environment with some sort of secret it would not reveal. I was haunted by it, and intrigued. Had I not moved away from the coast, I suppose in time I might have become obsessed.

As it is, I am pretty much of a casual beachcomber these days, walking along with my shoes in my hand, protecting my vulnerable bald head with a cap advertising Garcia fishing reels or my father-in-law's oil-jobbing company. I walk along the swash line, the ever-changing boundary where the played-out waves foam against the hard-packed sand. A carload of surfers slips past me – it's annoyingly legal to drive on the beach in Texas – and I'm treated to the brutal rhythms of Bon Jovi for a moment, as well as to the reek of Panama Jack. In front of one of the new condominiums a woman is walking an overfed possum on a leash. I do not ask why.

My eyes are trained ahead to the eternally hazy horizon. After a while the condominiums peter out, there are fewer cars, and the more or less unspoiled Padre Island assumes its rightful primacy. The island is never more than four miles wide, but it is infinitely long. In Boy Scouts we used to take punishing, industrious "nature hikes" up and down the beach,

trudging along through the sand with full packs, never seeming, in that blank expanse to progress or arrive anywhere.
Nature was never discussed much on these hikes. We were too
thirsty during the day, too tortured by sand fleas during the
night, to care about anything but our imminent deliverance.
But when my curiosity began to override my discomfort, I
began at last to pay attention.

For someone like me, a perpetual naturalist-in-training,
Padre Island is a perfect laboratory. At the beach, at the seam
of water and land, the world has a coherence for me that it has
nowhere else. I can see how the parts are joined together, and
I have confidence that observation will be rewarded by under-
standing. Every creature, every sort of natural process, seems
isolated for inspection. There is the track of the wind in the
slip faces and contours of the dunes, plain for all the world to
see. Drifting along through a tidal inlet is a sea hare, an ocean-
going slug as large as a baseball mitt. Willits, plovers, sander-
lings, and other sand-pipers dally with the waves; there are
gulls and terns overhead, and brown pelicans amble low
enough over the breakers for the spume to touch their wings.
You can feel the wind change direction against your face, and
then you can turn to the ocean and see it manifested in the col-
liding vectors of the waves. In the distance there is some tan-
talizing apparition: a long, gray parched shape that may be a
beached whale; a dark streamlined bird, perched on a drift-
wood log, that may be a peregrine falcon.

But there is more to the island than the beach.
Periodically there will be a hurricane pass, a wide corridor
that generations of storms have scoured through the dunes.
Typically, these passes open out into immense flats, miles of
hypnotically bare sand. At the edges of the flats are dunes covered with luxuriant grasses and wildflowers. And if you head
west across the sand you will come eventually to the leeward
side of the island, to salt marsh and mud and murky waters
thick with the larval beginnings of marine life.

You can still find shards of pottery, decorated with natural asphalt, that mark the places where Karankawa Indians once camped by the bay shores and slathered their bodies with alligator grease to repel mosquitoes. One this side of the island, scattered here and there in the grasslands, are brackish ponds where roseate spoonbills root through the mud.

This is the most isolated part of the island, and the most mysterious. Over the years I have learned to name most of the birds, and to understand them a little. I know the reptiles that live in the ponds and the mollusks that live in the tidal zones. In my haphazard way I am learning about sedges and grasses and saltwort. But the sum of these parts will, I know, always elude me. Padre Island will always be a brooding, unreachable presence for me. I am not drawn to it so much for its beauty as for some richness, some resonance, I can barely perceive. Here on the back side of the island, swatting mosquitoes and standing in black mud to my ankles, I have the feeling that Padre Island is the place where my education was meant to begin and end.

Carmen Tafolla

This River Here

This river here
is full of me and mine.
This river here
is full of you and yours.

Right here
(or maybe a little farther down)
my great-grandmother washed the dirt
out of her family's clothes,
soaking them, scrubbing them,
bringing them up
clean.

Right here
(or maybe a little farther down)
my grampa washed the sins
out of his congregation's souls,
baptizing them, scrubbing them,
bringing them up
clean.

Right here
(or maybe a little farther down)
my great-great grandma froze with fear
as she glimpsed,
between the lean, dark trees,
a lean, dark Indian peering at her.
She ran home screaming, "*Ay, los Indios!*
A'i vienen los I-i-indios!!"
as he threw pebbles at her,
laughing.

'Til one day she got mad
and stayed
and threw pebbles
right back at him!

After they got married,
they built their house right here
(or maybe a little farther down.)

Right here,
my father gathered
mesquite beans and wild berries
working with a passion
during the depression.
His eager sweat poured off
and mixed so easily
with the water of this river here.

Right here,
my mother cried in silence,
so far from her home,
sitting with her one brown suitcase,
a traveled trunk packed full with blessings,
and rolling tears of loneliness and longing
which mixed (again so easily)
with the currents of this river here.

Right here we'd pour out picnics,
and childhood's blood from
dirty scrapes on dirty knees,
and every generation's first-hand stories
of the weeping lady, La Llorona,
haunting the river every night,
crying "Ayyy, *mis hi-i-i-ijos!*" –
(It happened right here!)
The fear dripped off our skin
and the blood dripped off our scrapes

and they mixed with the river water,
right here.
Right here,
the stories and the stillness
of those gone before us
haunt us still,
now grown, our scrapes in different places,
the voices of those now dead
quieter,
but not too far away. . . .

Right here we were married,
you and I,
and the music filled the air,
danced in,
dipped in,
mixed in
with the river water.

   . . . dirt and sins,
   fear and anger,
   sweat and tears,
   love and music,
   blood.
   And memories. . . .

It was right here!
And right here we stand,
washing clean our memories,
baptizing our hearts,
gathering past and present,
dancing to the flow
we find
right here
or maybe –
a little farther
down.

A.C. Greene

Of Green Water and Invisible History

Texas is too big a land to make many generalized statements about, especially when it comes to environment. But individual problems can be attacked individually. I will name only two environmental disasters with which I have first hand knowledge.

First is the degradation of Texas streams and waterways. For instance, the Clear Fork of the Brazos River in West Texas runs absolutely green in many parts because of sewage disposal from cities along its course. This is usually "purified" sewage, true, but it also includes runoff from agricultural pesticides and fertilizers, indiscriminate dumping by individuals, businesses and institutions. Today, where my family once fished and camped for delightful days on the Clear Fork, one would not think of eating any fish caught from the same waters.

Unfortunately, this can be said of a great many other Texas waterways, although it seems to me that the problem might be remedied with a minimum of laws and maximum of public interpretations, focusing on the loss to everyone in a state where water availability and use have always been major factors in human existence in vast portions of the Lone Star State.

Second is the environmental problem that is seldom considered part of the environment: the loss and decay of so many frontier forts and deaths of towns and cities that were once centers of life and government. I do not propose to try and resurrect every lost town in Texas, but to preserve and reclaim such things that represent the Texas past which has given us modern Texas. In this respect, both Hispanic and Black history are being obliterated by changing public opinions. The frontier military forts, for example, are living history. It is so

much more satisfactory to visit such a place as Fort Griffin or
Fort Phantom Hill than it is to see a romantic rendition on television or in a novel. The human environment is certainly
enhanced by our recognition of our past, because the past
never really loses its effect on the present. Again, there should
be a few needed laws but also there should be supervision and
funds available for such public projects.

Bud Shrake

In the Valley of the Violet Crown

Once upon a time, not so very long ago – as a good fairy tale should begin – there was a place called Austin. "It is a pleasant city, clean and quiet, with wide rambling walks and elaborate public gardens and elegant old homes faintly ruined in the shadow of arching poplars. Occasionally through the trees . . . one can see the college tower and the Capitol building. On brilliant mornings the white sandstone of the tower and the Capitol's granite dome are joined for an instant, all pink and cream, catching the first light."

Thus was Austin described by Bill Brammer in his novel, *The Gay Place*.

But this is truly a fairy tale, alas – about an enchanted land that no longer exists.

Austin was chosen to be the second capital city of the Republic of Texas by President Mirabeau Lamar in large part because of the beauty of the river valley surrounded by hills – "like the seven hills of Rome," wrote Lamar. "The valley of the violet crown," he called it.

Thus Austin was founded to be a real estate development 162 years ago, as Houston had been founded to be a shipping center, and as Dallas would later be founded as a place to do commerce, and Fort Worth as a cow town.

In Austin they chopped down the trees and shaved the hills and destroyed the elegant old homes. They filled up the roads with automobiles, They poured concrete to cover the pastures. They let the parks go begging. For good measure they turned the air brown.

And as in a fairy tale, this Austin, this goose that laid so many golden eggs, is dying of a disfiguring disease. But let it be remembered that this fairy tale town furnished much pleasure for many, before we turned on it and killed it.

Michael Adams

How to Kill a Salamander: A Bedtime Story

Once upon a time in Hays County, Texas, 350 residents whose wells were running dry requested help from the Lower Colorado River Authority (LCRA). The men who run the LCRA, all political appointees of the Governor, are also financial friends of the developers who see the rare beauty of the hill country as nature's façade for twentieth-century gold. Help? Yes, indeed. Under its statutory authority, these political friends of the developers hastily authorize a regional water system to serve over 76,000 connections in Hays County – the lifeblood of development. Little attention had been paid to the fact that they completely ignored the Endangered Species Act, which demands that they, of all things, protect endangered species.

The consequences of development like this in Hays County has had a direct impact on one of Austin's most famous jewels – the 1000-foot-long, spring-fed, 68-degree Barton Springs, long sacred to the local Native Americans and current sacred monument to those in the environmental movement, seeing it as they do as the ultimate battleground between beauty and profit, value and price, a way of life and another McDonalds. Those who took up this battle are acronymed SOS – Save Our Springs – and they have made it their mission to fight against the violation of their way of life. And their way of life is not to have Barton Springs periodically closed because of toxic levels of pollutants and fecal matter. They make the battleground the Court, in this case filing suit under the Endangered Species Act, since development upon the Edwards Aquifer, which feeds the springs, is also threatening the extinction of a unique salamander that exists only in this rare place – a symbol not just for life, but a special life.

The story of this life, symbolic and real, is the suffocation of the salamander by development run-off, petroleum hydrocarbons, heavy metals, and high concentrations of pesticides. It takes us behind doors where politics and money, as always, become the alchemy for power – for very special interests, in this case the developer who oxymoronically calls his idealism "city in the country." It is an ideal that always morphs into an urban congestion of walls and marts and gasoline fumes. And the story takes us into the lucrative world of political barter, euphemistically called campaign contributions in exchange for votes from local city, county, and state leaders. What does the developer want? For Austin environmentalists to stop messing with his messing up of Texas.

The salamander's story is replete with amazingly flagrant developer victories: the persuasion of a state representative to sponsor a bill that would create just for him, the developer, a special Hays County taxing district that would allow this special district, controlled by the developer, to borrow money, annex land, and condemn property; the persuasion of Secretary of the Interior Bruce Babbitt to remove the proposed listing of the Barton Springs Salamander as an endangered species. Federal Judge Lucius Bunton found that Secretary of the Interior Bruce Babbitt had violated the endangered Species Act, concluding that Babbitt's decision came directly from "strong political pressure" from "the political lobbyists," which translates as developers' hired guns. Alas, this is the essence of the story.

It is a simple and sad truth, this story. Greed, Developers, Money, Politicians win. It's the driving purpose of each character in this economic game-world. Yet, SOS has proved that it can win too. But can they sustain their periodic victories? The multi-billion dollar Lowe's Corporation is maneuvering to build a 160,000 square foot behemoth over the Edwards Aquifer, not just the source of Barton Springs, but, as SOS has underscored, the only source of drinking water for thousands of Central Texans. Watchdogs like *The Austin*

Chronicle and SOS have already exposed the secret dealings between the LCRA and developers that will inevitably destroy the beauty, pollute the water, and ruin the quality of life for those who cherish the hill country.

Unfortunately, developers will never run the risk of going extinct. We will never need federal legislation to protect them. They are the cockroaches of the economic and political ecosystem. Perhaps that's too kind. They are more a hybrid, part predator, part insect, part alpha dog, part Cro-Magnon. The death of a sacred spring stirs no emotion in them, for they have already bought their pristine hideaway where they can enjoy their peace and quiet, their clean water and running brook, their fresh air and silent sunsets, and enjoy their next plan for transforming someone else's haven into lots of cold cash – where they can read their children to sleep with: "Once upon a time there was some beautiful land not yet developed."

(For more extensive accounts of the struggle to save Barton Springs, see www.sosalliance.org/www.austinchronicle.com)

Bill Crider

The Pool

The pool is rimmed with iridescence,
Red and green,
And slick-veined moss,
Like mucous to the touch.
On the bottom there is, perhaps,
The rusting fender of a '54 Ford
Or a rotting synthetic tire.

I can imagine froggy fingers,
Cold as stone,
Sinking through mephitic ooze,
While bulbous eyes stare upward
Through wavering, sickly light.

This pool lies precisely as far
From Walden Pond
As Thoreau's life lies
From mine

Michael Berryhill

Most Evenings

I've been reading the stories about how the Taliban rulers have destroyed the huge stone sculptures of the Buddha carved into the side of a mountain. There was an international outcry about the destruction of these cultural artifacts, and condemnation all around. Although a Buddhist myself, I didn't lose much sleep about it. So many sacred things are being destroyed deliberately.

I write this from my house in the fishing village of Seadrift on the shores of San Antonio Bay, in the middle of the Texas coast. The next bay up, Lavaca Bay, contains an underwater Superfund site. Despite the lessons of Minimata (remember those terrible pictures of the dying children in *Life* magazine in the late 50s?), Alcoa was allowed to dump mercury into the bay. Now they're spending millions trying to figure out how to get it out, and they're not close to a solution. There's a warning on the causeway not to eat the fish from that bay, but I see people fishing every time I drive by, which is often twice a week. And I don't think these fishermen are into catch and release. The fetus of a pregnant woman could be irrevocably damaged by eating fish from this bay.

South of me Dupont Chemical and other companies are pouring waste into the Guadalupe river, which flows into San Antonio Bay. Dupont applied and won a permit from the state of Texas to put tons more chemicals into the bay, with a poorly thought out plan to clean up its waste water.

Every day our industries and cities pour pollutants into Texas bays. We regard the Taliban as slightly barbaric and self-centered for destroying ancient statutes, and we continue to poison our bays, not because of some strongly held religious belief, but because we're too cheap to clean up our waste stream.

Of all the western states, Texas is least in the amount of public lands, stemming from its history as an independent republic. We can't count on the federal government to hold our natural (and spiritual) heritage for us. In Texas our public land is our system of bays, under-appreciated, stressed, dug up, poisoned, and yet somehow, beautiful.

For the last year I've been traveling up and down the coast, working on a project called The Book of Texas Bays, which is being underwritten by a major grant from the Houston Endowment and several other Texas foundations. My partners in this project are environmental attorney Jim Blackburn and photographer Jim Olive, who have been involved in the coast far longer than I.

I've had my eyes opened to the richness of the coast. Our bays are as productive as any rain forest. They are amazingly resilient. Galveston Bay, situated near four million people, still produces some of the Texas oyster crop. Our blue crabs are shipped to Maryland because the Chesapeake Bay has become so damaged it can no longer keep up with the demand. The Laguna Madre is regarded by the Nature Conservancy as one of the top twenty eco-systems needing preservation in the world. The Texas coast, 367 miles in a straight line, but thousands of miles long if you count the intricate coast lines of the shores and barrier islands, is our wilderness treasure, accessible to millions of people.

Like many Houstonians, I have been a weekender to the coast. I built a house and ended up moving here full time, enrolling my little girl, Elizabeth, in the local school. People still fish for a living here, harvesting shrimp and oysters. You can eat the fresh product of the bay at the local cafe here. How much longer this will last, I don't know.

Sports fishermen want to drive the bay shrimpers out of the bay. They have the money to do it. Instead of fighting the industry and the urban pollution that are harming the bays, they go after the shrimpers, the easiest, poorest target. We scorn the Taliban for destroying sculptures from another religion. How much better are we in destroying a whole way of life that brings wonderful wild foods to our tables? Of course the shrimpers will probably be allowed to trawl for bait.

Because money rules. We could clean up these bays. But it's going to take not only political pressure, but I think, a sense of shame. The chemical industries could make a goal of zero discharge. Does that sound so incredible? The goal was written into the original clean water act in the 1970s, but was quickly taken out.

Look how far we have come when we do take action. Out in Mesquite Bay is a group of isolated islands called the Second Chain, where the brown pelican and roseate spoonbills made their recovery during the last 30 years. The southeast winds are blowing hard now, and I love it, for I go windsurfing out in the bay at blazing speeds. I go far out in the bay past a bird island. Egrets have already started nesting there. As I make my jib and turn back to shore, five roseate spoonbills, pink and crimson from eating shrimp all season, fly past me, headed for the island. I will watch this little strip of shrubby land through a telescope from my balcony all summer long. It is amazing how productive it will be: probably close to a hundred nests will crowd this little place.

The water is whipped up to a lathery, olive drab, not the color that draws the tourists. They will drive hundreds of miles to the sugar sand beaches and gin-clear water of Florida. But this water, sometimes the color of coffee with cream, is full of nutrients because rivers feed into it. It is the color of life.

Living close to a wilderness changes a person, makes him look, look all the more closely. There are wolfberries growing around here, a kind of small red pepper that is a favorite food of the whooping cranes who winter here. One morning I had to take Elizabeth late to school, and a gray fox ran across the road and into the brush. A dozen miles from here is the northern most example of Tamaulipan brush, stuff so dense and thorny it's opaque. It's only 70 acres, but it's a huge resource for migrating neotropical songbirds if they hit a norther and arrive exhausted from the 600-mile journey over the Gulf of Mexico. Local birders are trying to save it from development.

I don't need to say much more about what it means to be near these birds and animals and plants. Everyday they teach me something more, connect me to what it means to be alive. The other day I sat under the one-lane bridge and watched the swallows that have recently come back to breed. The books tell me they came all the way from Argentina. How near at hand! Closer than the Taliban.

So this is how life is going. Here's a poem I wrote one evening. Elizabeth speaks for me:

Most Evenings

Most evenings, Elizabeth wants a walk
along the bay front, and so we go. Binoculars
around my neck, her yellow tennis ball
stuffed in my pocket. At the pavilion
we stop and play catch. The western clouds
are marled orange and pink. White pelicans glide
on the molten blue water. Three skimmers
swoop in, shearing the fabric of the surface.
We catch and throw, catch and throw until
Elizabeth finally sits down and folds her hands
in her lap. "Are you tired?" I ask.
"No," she says. "I'm happy!"

Robin Doughty

Ponds

Treatment Lagoon (One West), Hornsby Bend
Biosolids Management Plant, City of Austin, Water
& Wastewater Utility, F. M. 973, Austin, Texas.

In the first chill I hold peeps
marbled in midtwist by
wheeling light,
pearls against aqueous lace,
dun in a succulence of mud.

Names stare from the page,
declare rank – long billed or short,
white on brown, shanks in
different sizes. Book poses,
dull and witless.

These wade-wings stretch and
teeter. Their needle beaks tap
soft earth, pluck worms, insects too
small to see except by ruffled
jousters.

Into a creche they press and loaf.
Each slicks feathers, and every
feather slicked positioned
into place, glistens for the
travail-to-come.

Startled. They fling their feathers
as one hawk steadies, twist higher
in tumult as another threads blue,

scouring wet places for these
beads of energy.

Guarded by infirm edges,
fueling for migration and a north wind,
as far as Santiago, even where penguins live.
They linger here in fall,
and shimmer.

Robin Doughty

China Round

(Barton Springs, Austin Texas)

In slow light, the black bough
lifts above the Pool.
Photo in black and white,
yin-yang in a Chinese
painting. Life stilled. Waiting

for human ripple, first
cut of my hand
like a knife
through limpid
quiddity.

Under I crawl,
like a prehistoric squib,
bucking my way on
spider arms and legs
toward the tree.

I must honor this bough –
all still simplicity. Raccoons
shuffle, geese honk
in a nearby creek.
Aloft, a heron kronks.

The bough lies black and silent.
I splatter the surface in a sudden
drizzle, like shook fur. Salutation.
Then, with a heron's rasp,
I dive.

Fern – the bough's water-blessed
replica – looms, a netherworld
yang where fishes work.
And new gills begin
to pump.

Oxygen makes my scales glitter,
fins speed me
as I shift from man to fish
through Paleozoic
re-awakening.

Slanting down,
I swim on and on,
toward the amber
gates of
China.

Paul Christensen

Water

Murray is my broker. I call him up occasionally to talk stocks, and once in a while I make an appointment to see him at his office. I don't like his office; if there were some other place to meet him, a bar or a park bench, somewhere warm and personal, I would. Believe me, if there were a way to avoid seeing Murray at all, I would.

But the problem is my dwindling money. I am slowly losing my investments as the market goes dry. I hate the stock market. It is not a real world. It is something you would expect to happen from the spread of Protestantism. A stock market in which God is a quotation board and his divine will is only the flicker of numbers going up and down.

Is it an accident that brokers all look like Presbyterian ministers? Or that brokerage houses, I mean the small ones in backwater towns like Bryan, Texas, where I live, have the atmosphere of chapels? Murray reminds me of an old country parson with a congregation of farm couples and their cross-eyed sons sitting in the heat while Murray swayed over his lectern airing one of his speculations on the mystery of money.

So I am once more calling Murray for his advice about my dwindling fortune. I am not a rich man, only an underpaid professor at the university. I was never concerned about stocks until my mother died and a small portion of my father's estate was settled on me. Murray advised me to sell the blue chips and transfer out of some slow mutual funds into what he called "aggressive" accounts. I did and that's when my losses began.

"Hello, Murray?"

"Yes," he says, "Is that you, Paul?

"Are you free tomorrow around three or four?"

"Let me look." Longish pause. Receiver crackles. "Four's good."

"Fine. See you then."

—•—•—•—

Murray's office is at the end of a crosstown feeder that dumps traffic onto the southbound freeway to Houston. There's a hot little parking lot full of glinting car roofs and a flag stone path that goes by other brokerage offices to his, which is tucked away under a rusty outdoor staircase. You go in and find the receptionist window empty. He's never hired any staff. The rug sets off a door chime and you wait until Murray scoots out into the hall on his chair and says "C'mon down."

"C'mon down," he tells me.

I go down the dim hallway to his office, first on the right. It is paneled in fake wood that has bowed out from the walls. Acoustic tile on the ceiling, long tube lights that give Murray's pudgy face a deathly pallor. He has bulging, soft eyes and a gut that pulls his shirt buttons tight. His gray jacket hangs on the coat rack next to his umbrella, which he has never used.

He's been punching the keys of his computer and a spread sheet glows on the monitor. Another computer clicks up the half page of my investment and sits there waiting further instruction. He scoots in behind his desk, which is littered with the usual brochures, note pads, pen stand, phone with lights blinking. A small window looks onto a blazing strip of concrete where the trash cans are lined up. He pulls the blinds shut and we go into ether light.

"Got any tips?"

"Not much," he says in his sleepy nasal voice.

Murray is different from ordinary people. There is something strange inside him, under that large, balding crown, behind those rubbery soft eyes and lazy mouth. He's not the brightest man you'll ever meet or the shrewdest. He's hardly what you would call a success. He lives simply, has a large fam-

ily whose grinning faces adorn his book case. He's basically a
simple guy with a humdrum life.

But he was born with the nose of a truffle pig. He's all
snout when it comes to sniffing out the hidden profits that lie
under the strange landscape he calls the financial world. But
what Murray calls "profit" has hardly any resemblance to the
green stuff you spend. Profit to Murray is the riddles and
paradoxes of a greedy culture, the wasted money that pools in
the underground like a trapped spring. Even in a desert,
Murray will say, water is just beneath the burning sand, the
barren treeless wastes where the dumb brutes lay dying of
thirst. You only need to know where it lies to become well off.
Murray hates the word rich; I've never heard him use it. You
don't become *rich* because that is merely another pool of profit lying trapped inside a man's pocket.

According to Murray's philosophy, a man or woman
should only want to take the money that lies fallow in the world,
that no one else knew about or cared enough to put to work. He
believes everyone could be happy in America if they took their
humble portion of the money available and spent it wisely,
invested in good works, and gave the rest away to those getting
started. The sins of America, the cause of its decay and tragedy,
are greed and ignorance. Wasting money was like pumping out
water to feed a cotton field when there was a cotton glut. Or like
growing sugar cane in Cuba and nothing else, so that *you would
definitely bring on a revolution!* Murray liked to end his logic
lessons with a sort of aspirated serpent hiss.

The day I came for my visit, Murray was in high spirits.
As usual he had "called up" my account, which hadn't
improved much since he last called it up.

"It's a sick portfolio," he said. "It's not responding. We
need some radical surgery at this point."

"What's happening to my money," I asked darkly.

"Nothing. Nothing. The big mutuals are tied up in cigarette stocks which they're trying to dump. All the prices keep
coming down down down."

"We gotta get out, right?"

"Right," he said. He was thinking to himself.

"So how," I said, trying to be helpful.

"Ain't so easy. I don't see any movement on the board. It's a crazy time. People are saying hang tough, but it doesn't seem right. I mean money and jobs are going to China and won't come back. Iraq is a hole that can't be filled with our loot. It keeps eating it and wanting more. It's gonna be a long time before anything gets interesting. I have a feeling," he said, rubbing his elbow.

"Okay, Murray. What've you got?"

"It's just a feeling, okay? But what I see out there is a big drought, I don't mean a money drought, that's already happened. I mean a real drought, a drought on this side. You know, real water. The stuff you read in the papers nowadays is all about dropping water tables, farmers walking away from cotton and corn because they can't afford the water bills. They keep sinking new wells in Texas but the stuff is salt, shale oil, chalk, zip. So."

"So?"

"So, I say liquidate your assets and let's do something totally wiggy."

"What is wiggy?"

He squished around in his chair. What he wanted to say next took some preparation. He wet his lips and tried to think of how to get into the topic without having me bolt from the chair.

"You ever think about religion," he asked shyly.

"I haven't lately," I said.

"I do, all the time. If you live in the bible belt, you think about it a lot. I mean, I can't escape it. You go off and travel and give lectures and things, but I sit here all day long talking to the other agents. They all believe in something or other. My wife's religious. The kids go to a little private school where the teachers all talk religion. You can't escape it. Texas is a very religious place," he said. He scratched his cheek and looked up at me like a hound dog.

"I don't follow."

"The future is water, my friend." Long sigh. Some business at the desk with his pudgy fingers. He knew how to stage his big moment. We sat there in the white noise without moving. "Water will be all there is of money one day. The oceans are dying, the rivers are poisoned, the wells are going to salt. He who hath water lives and prospers. He who can tell when the Nile riseth will fill his silos early and stay a famine."

"I'm sorry; I'm being very dense, but all this is going right over me. I don't see what you're getting at about religion, water, salty wells."

He looked tired. He sat in his swivel chair and seemed to go all to fat. I thought queasily that my own *weltschmerz* had finally infected him. Maybe there was a soggy bottom to Murray as well. Old solid Murray was bubbling up from the doubtful ooze.

"Look," he said softly. "Here's a map of Paris. Do you see anything funny about it?" He held it up so I could scan over it. It looked perfectly ordinary to me as a map.

"It's very nice," I said.

"It's quite old. It was done by a Jesuit priest in 1838. He located every chapel, church, basilica, shrine, cathedral and cemetery he could and put them all onto his map. He did it for some bishop who wanted to put up a new church and didn't want to crowd out somebody else. So here it is, the whole Catholic system laid out, street by street." He looked over to see if I were still in suspense. I was.

"And this," he said, pulling out another map from his mound of papers on the desk, "this is a modern hydrology map of the same Paris." He held it up so I could examine it.

"See anything unusual?"

"Lots of water," I said carefully.

"Lots of water, lots and lots of water. Some from the Seine, some from little underground streams, some from springs, wells, pools, some from seepage, all the water that lives under a city. And now look." A brief smile glimmered and died on his lips. He was moving in for the kill. "Look at my church map and tell me if you see any relation between this one and the water map."

I looked and I could see that some of the larger churches were near the Seine. I told him so.

"Yes, but you haven't gone far enough, mister. There ain't one little bitty house of God that ain't square over a water hole!"

I took another look. Sure enough, the slightest little monopoly house marked on the map with a cross stood directly over some part of the blue veins of Paris. It was uncanny. The more I looked, the more it came clear that *every* bit of the machinery of Catholicism was built over water.

"Water *is* God," Murray whispered close to my face. "Christianity ain't nothing more than a water religion. There's no building ever put up in the name of that faith that didn't have some relation to the wet stuff. Baptism by water, communion with water and wine, washing of the feet, Mary's blue dress, the sky as heaven, you name it, it's all water." He folded up his maps with a gloating look.

"Well, I'll be damned."

"Damnation's fire, the opposite of water," he said in his smug preacher's voice. "And the world is headed for damnation, a death by fire, a drought that will squeeze every throat but the faithful. You don't think those Mormons knew something when they pitched their little tent on Salt Lake? They were saying, you don't join us you can drink this old pond of damnation and die. Join us and you can chant with the choir and drink out of the white salamander's own private swim hole."

He had me, I admit. But he wasn't through winning me over. That's the genius of a great salesman. They don't quit when you start smiling. That's only the beginning.

"The pope is called the pontifex maximus, the greatest bridge builder," he said wearily. "So what do you build a bridge *over?*"

"Water," I said, like a good front-row student.

"Exactly. Water. Now from a satellite the Earth looks like a big water ball. It's the only stellar body we know that is mainly water. Nothing like it any other part of the universe. Does it not stand to reason that this water from which all life is made and all things depend, this unique substance in the universe, is not in fact God himself? Then any temple anywhere on this blue ball is a shrine to water. A house of worship is a waterhouse, a roof over a well with a chair to pray in. And salvation, my friend, salvation in the very near future will have one meaning and only one, water. Sin and evil are the poisoning of water that will make most folks die of thirst. The saved will be those who pray *over the water* in the house of God."

"Amazing," I said. "But Murray," I stammered.

"You want to know what this has to do with your money? Of course you do. All money is shrinking now because it is invested in the poisoning of the great blue ball. Secretly, unknowingly, we are revolting against the further contamination of our planet and our dread takes the form of dying markets. Capitalism has gone down the wrong path; it has outgrown its own religious justification and offends the very thing that mothered it. Money is being put into the death of nature. The only hope left is in religion, in its water God. He who owns the water rights below the churches of America owns the only life-sustaining thing left on the poisoned ground."

I got it.

⊱──◦─◦──⊰

We liquidated every thing I had; it was all cash now. Murray put some of it into CDs and the rest he parked in a savings account. We went out riding in the countryside to look over some of the lesser establishments of the Lord. We brought along a man named Smiley Higgum, a notorious wino from the Baptist Mission, who sometimes did my lawn for me and swore he had been a professional dowser in his salad days. He had a branch of hickory he was skinning with his draw knife. The wood smelled cool and minty in the car. I liked the whoosh of the blade as he worked, chewing his tongue with each pull of the knife. Murray spoke in code the whole way out to Snook, which was our quarry this early, hot morning.

The streets of this desperate little town were all burned to a crust; the tractor dealer had closed up last winter and the post office was on reduced time. The hardware had been turned into a bingo hall; a big Furrows stood on the hill of some old farmer's field turned to asphalt. Here and there a shack or cottage threw out a ribbon of shade onto the brown grass. A dog napped in the full sun of an empty vegetable stand.

We stopped in at the sheriff's office to tell him our business, which was to write an article about local ministers for a Baptist magazine called *Bright Day*. He bought it and said we could find the Reverend Lester Swanson down at the gas station. We found Swanson sitting on a pile of new tires while his oil was being changed. We made a few pleasantries and got down to business. Swanson said we could poke our heads into the First Baptist Church of Snook and even fished out the long silver key to open the door. He was very happy to meet us. Smiley stayed in the car to work on his hickory branch.

The First Baptist Church was a small building with a steep roof and a row of white columns along the porch. The door was painted a blinding white and the porch was carpeted with Astroturf and glowed like neon. We opened the door and a cool, dry odor of shellac and bibles wafted over us. The room was plain and orderly with a bare stage and lectern, a few flowers, and a flag pole. The good book rested on a window sill with markers all through it. We drummed over the shiny wooden floor to the altar and stood there as Smiley came through the front door holding his dowsing rod.

Smiley wore a fierce expression as he crept slowly up the aisle toward us. His eyes were slitted and his mouth kept working

with each step he took. He was on his wages, he kept saying on
the way over, and he worked hard. He moved like a praying
mantis over the long glittering floor, pulling up his foot and
setting it down very carefully then chewing his tongue hard.
The branch was rock steady in his hands.

Murray sweated profusely and mopped his brow with a
red handkerchief. He was all for going to the supermarket for
a cool drink, but we were compelled to let Smiley do his work.
It would be a while before he got up to us. The long walk
required all his attention and our complete silence. The noon
traffic murmured by, a few pick ups and an old blue tractor.
The Reverend Swanson had gone home to lunch and would
meet us at 12:30, he said.

Just where the step began Smiley let out a gasp. "It's
workin," he said tensely. The hickory stick, a little fork of
wood he had shaved down to the white wood, trembled a little.
Smiley was in a trance and his foot came up very slowly and
landed like a butterfly on the step. The hickory pulled down a
little. Then Smiley stood up fully on the step and the hickory
dipped down and slapped Smiley on the fly of his blue jeans.

"There's more fucking water down there than there is in
the whole goddam ocean!" he declared in his official capacity.

"Murray!" I said, whirling around to find him beaming
with pride.

"I knew we shouldn't sink our telephone cable under
here," he said, winking broadly to me.

"Cable?" Smiley asked, foot in the air and eyes shut
tight, the wand dipping and flapping as he slid forward toward
the altar. "It's a river, by god," he said. "I'm crossing it now.
Deep little fucker, 'bout ten feet or more. Middle of it now, see
my stick?" It curved down from his grip and almost seemed to
break from the force. "Deep, deep river, this little gal.
Running fast, real fast. Here she is coming up again, getting
over it. Yeah, right 'bout here I'm on land again," he said. The
stick rose straight again and was loose in his grip.

"Which way she running," Murray asked.

"East west," Smiley said, "way they always run in churches."

"Always?" I asked.

"Yeah, all churches got water under the altars," Smiley said, moving slowly along the stage toward the window. "It's a good little river, this baby. Yeah, she's fine. Full of life. Yeah, all the churches I ever been in have water under 'em. Lots of it. It's a fact but no one'd ever believe me when I told them. I mean, they say, Smiley, you some fuckin' ole wino, how you know shit? Well, I reckon I know shit and more. I know water, I know where all the water in Texas hides. Ain't no secrets water can keep from me, or my pap or his pap on back to Adam. And there ain't no church that don't have a mess of water under it. Don't ax me why, zactly, but they all do."

We made a deal with Reverend Swanson and he was delighted to sign over any water rights he thought the church possessed for the mere sum of two hundred dollars. I wrote out a check and he signed the document Murray had drawn up. That took care of Snook's water supply. We marked it off on our map with a large W.

⊢•◦•⊣

We went to Gause, next, a moribund row of clapboard buildings, most of them abandoned. A siding held a few grain cars from the previous winter, and a cement mill lingered on the edges of the new road. All in all, twenty people kept Gause running. But the little chapel called the Good Faith Temple wore a new coat of paint and a fine, gray shingled steeple. A circuit rider came around three Sundays of the month and a lay preacher took the fourth. We got the key from the cement company and found the place musty. The dust was thick as felt on the pews and the little varnished table up front. There were padlocks on the windows and a few broken panes of glass had been repaired with cardboard curling off the tacks. A large blue spider hung in the space where the bible was kept.

Someone had lugged it off to the laundromat for safekeeping, we found out.

Smiley came up the aisle in his slow, insect pace with the dowsing rod held out rigid. Two little black boys peeked through the door at him until we ran them off. Smiley got all the way to the table in front and then over to the corners without a sign. He walked up the side aisle and down the center. Nothing. Murray grew restless. He looked at the hickory branch and gave it back to Smiley with a shake of his head. Smiley didn't like it either.

"What you boys want in here," a woman said, a large black woman with a basket on her hip. She was very forceful and serious.

"We're doing an article on old churches," Murray said. "For *Bright Day*, ma'am."

"You got a dowser, you lookin' for water?" She'd heard from her kids.

"Just to check on some things, ma'am," Smiley said, putting the branch behind him where he sat.

"Ain't no water here. It's out there," she pointed west. "Where the old church was."

"Thanks," I said, "that's fine."

When she left, we waited a moment or two and then walked over to the old church. There were remains of a few brick pillars where the floor had been. A cat sprang out of the bushes and ran off. Murray figured out where the altar was and Smiley walked up where he thought the aisle began. When he got to Murray the rod bent down and slapped his thigh and then curved over and creaked as Smiley held onto it with white knuckles.

"She's down there, boys. Half a damn sea's in there," he said through clenched teeth.

"Pay dirt," Murray said, forgetting code.

We signed up Gause on our water map and went off. I paid fifty dollars for the rights. Mrs. Oldchamps took the money in the name of the community, as she put it. She put the check under the rabbit ears on her TV and that was that. She was the widow of the last mayor Gause elected, some twenty-seven years ago. Seems she had power of attorney for the town, or so she claimed. Murray would check on it.

Six months went by and we had signed up some forty-three water rights from the churches within a fifty mile radius of Bryan. Murray said we had the makings of a water company and had cornered the wet market of central Texas for the benefit of mankind. We were following in the footsteps of the oil companies and the gas men and the hard rock miners and now it was our turn to discover a precious mineral in the ground. I had spent more than half of my money on this venture and was still dubious about the outcome.

I asked Murray to explain what I was doing again. I had fallen into another bog of doubt and couldn't quite climb out. He knew me well enough to be patient.

"You ain't hoarding the water, mister," he began calmly, with great self-assurance. "All we're doing is forming a water company. I will buy some of your rights from you and balance off your risk. That way we will have a partnership. When we have got all the water rights we can afford, we will approach other investors and go public with our company. Plain as that. When the word gets out that we have the last remaining supply of fresh, drinkable water in this part of the state, there'll be a bum's rush to buy up the rest of the water from here to the Pecos. The churches that still own their water will drive up their market price. Our stock will double, triple, and go crazy. We'll wait and see how the market does; when she's high as she can go, we'll sell off a bit; if she holds steady, we'll sell more. We'll freeze fifty-one percent of our stock and keep voting majority in the quarterly meetings. But you and I will be very comfortably set. You can count on it," he said, relishing each syllable of that last remark.

By the time Murray and I were through checking out the churches of Texas, we found a lazy, broken down, indifferent religion, a careless, fool-hearty attitude to God. These were the minds that threw up chapels and churches over sewers and mains and ordinary water pipes, and missed the big streams for the trickle. They just didn't know. They just didn't give a damn. They were more interested in themselves than in water, in God.

You might say we had bought God from them when we got the water rights. And they sold it, as one preacher said to me, to buy a new fan. The faithful won't come to worship in summer without a swamp cooler or a good fan, he said. So here's God for a mess of air beaters.

Murray behaved like some conspiring dark angel in the wars of heaven. If I had doubted his good nature, as I did at times, it crossed my mind that Murray was not at all interested in buying up the water pockets under Texas, but in owning the very God for whom the white Protestants had fought the Mexicans and Indians and the bad weather to call their own. It crossed my mind, it ran in the stream of my consciousness, it battered the pylons holding up the floor of my own being, that this God whom Murray and I had purchased, might not belong to us.

⊢⊷◦⊷⊣

It was then I decided to bring out a well digger and perforate, as he described his work, the ground under one such dilapidated chapel whose water rights I owned. He hauled out his digger on the back of a truck and set up to make his hole. Some of the town of North Zulch came out to see the well digger do his work. We moved the old chapel to the left several hundred feet and cordoned off the work site with bright orange plastic fences. The well bit was reared to upright position and set down into the hard pan to dig. It raised a little dust and ground its teeth with much rattling of the gear box, and then began to spin off into the underworld, coiling up long strips of dark clay through its grooves.

Smiley elected not to come on this mission, and Murray was off on business in Dallas. I stood alone with Maurice Gunner, the well man, as he watched gauges and wiped a greasy towel over his forehead as the bit disappeared into the ground. He put on more pipe and righted the angle somewhat, and lit a cigarette. He eyed me a long time in the racket his digger made and then leaned over as if to kiss me.

"You ain't got a drop of water down there, mister. You're wasting good money," he yelled.

"I have records that show there is a river down there," I said self-righteously.

"Records, shit," he merely replied. "Bunch of old rocks is all, and it's gonna eat up my bit."

Just as he said so, the bit ate into something indigestible and the whole whirling mass of iron came to a halt. The clutch on the powerdrive slipped out and now the winch motor ran up into a high screech.

"Ain't a fucking drop of water down there, son," he said, throwing levers and crawling up the rigging. "Not one fucking measly drop of spit. It's your money," and he came down again, threw a lever forward and off the rig went again, turning and grinding and shaking hard.

Townspeople now crowded up to the orange fence and began whispering and looking.

"That some kind of oil rig," one old rancher asked, moving his Adam's apple up and down a raw, razor-nicked throat.

""No, it's a water well," I said.

"A which-its?" he asked, suppressing what appeared to be the start of a long, southern grin.

"A WATER WELL!" I shouted back. The whole crowd went into a huddle and the old man stared at me with watery blue eyes as if I had said something in Mandarin.

"A wa–, a wa-ter well? A well?" he asked, removing his sweaty hat and rubbing his wrinkled head. "Darn if I know'd there was any water here, for sure," he said half to himself.

"Wa – ??"

"WATER WELL, yes, a WATER WELL. That's what we're here to dig, right Maurice?"

Maurice chose that moment to blow his nose with the towel, an operation that took all of his attention away from the outside world.

I heard the word water pass through the crowd like wild gossip. The last to hear the word looked up with saucer-like eyes and mouth gaped open to behold a miracle in his midst. He stared as if a hornet had stung his brain. Water, he seemed to say, trying to speak and swallow in the same moment. Here?

Just then the tone of the rig went down an octave and the spinning slowed to glue.

"Christ awmighty," Maurice cried, shoving his towel into his jeans and running round to the truck again. "Feared that," he said, looking at me as if I were the devil. "Goddam, go and stick yourself in a fucking pile of granite shit, will you," and he kicked the winch motor hard with his boot. "Damn bitch," he said, climbed down and ran for his tool box.

"What's happening," I asked.

"What the fuck is happening. Mister, you'd be better off trying to stick this here drill bit up the moon's ass hopin' for water. Ain't nothing here but cow shit and clay and chalk at the bottom. Why you think they charge so much for water, I mean real water? Ain't no free water running around Texas."

"He's lookin' for water," a skinny young man said in a loud voice to no one in particular. He was gazing out into the fields when he said it. It was as if a prayer had escaped from him in a vision. He saw something out there, or nothing. But he spoke straight out into the void.

"Won't find it, neither," said the large, panting woman behind him, a beauty-salon customer called out in an emergency. She had her hair in a net and her fingers wrapped up in gauze.

"Could be, though," the minister said, the Reverend Axel Hazard, with whom I had done business some four months before. I had arranged for the church's removal to the side and for the digging to last a week or so. He vigorously agreed with me and asked no questions. He seemed to think I worked for some large company or for a foreign country buying up minerals in America. Perhaps he thought I was a Russian spy and didn't want to cause any trouble.

The drill bit into various layers of hard rock and sand and lower down some Austin chalk, then some brine and the traces of an old sea bed, some volcanic soot and a layer or two of thin quartz. Then the heart of the planet began, the secret inner life of the water house of God. The drill had sung many songs going down into the great body of the Earth. It had chanted strange mantras and Indian medicine songs, and one or two melodies of its own. Now it merely hummed a low, C major Om sound with occasional wavers. It threw us all into a spell hearing such soothing earth music. We stopped talking. It almost seemed the crowd had come over to my side and wanted water to shoot up the pipe and fall down in a muddy downpour. It wanted to see God come up pure and dark and powerful out of the hole. It would have underwritten all their unconscious mythology and faith if water had exploded from the pipe.

But the ground was a bloodless fossil that morning.

I called Murray and told him to get over in the morning. I would stay in North Zulch overnight and keep the operation going until something happened. Smiley had vanished. He jumped a freight for El Paso that morning, the guy at the Mission desk told me. He wouldn't be back until late spring.

———

As it happened, the hotel in North Zulch is owned by the minister's wife's family, the Krasnabeks, from the old Czech settlement in Caldwell up the road. Hazard's sister-in-law, Rosellen Krasnabek, had fallen off the straight and narrow a few times and was a hard country woman with angry eyes and a large, pouting mouth which she slathered with fiery lipstick.

She signed me in and gave me the room at the top of the stairs, next to hers, she said.

It was a dark little chamber with a rickety iron bed and a leaning dresser with one drawer missing. The floor creaked and the splintered pine boards had been covered once in linoleum in better times. A faucet hung down over a tiny chipped sink. The plug was long gone. Even the trap in the pipe was gone, and the water trickled out and puddled in its accustomed dent in the wood below. I heard a cough from my neighbor below to indicate the water was leaking. I shut the faucet and lay on my iron bed with the sun beating on my knees.

Why did I believe water lay in such cursed and abused ground? What good was it even if there were water in the hole? Only Murray seemed to know what we were doing with all this wheeling and dealing over a sleeping muddy God. I felt miserable. I shut my eyes. I heard my door push open with a rasp and I kept my eyes shut. I feared what I would see if I opened them. A gun? A minister with an axe over my head? A woman with a bible? I heard the sigh of silk dropping to the floor and the creak of shoes stepping out of cotton and snapping things and the press of some heavy, bony weight falling upon me full length.

When I did open my eyes, it was Rosellen spread-legged over me, her tongue going down my gullet with more tenacity than a drill bit. She knew where the watery Gods were hidden. I protested not. I pulled my shirt off from under her, and got free of my trousers and boots and found myself in the rollicking postures of a man in love.

⊱──━━━⊰

Murray didn't make it to North Zulch. It turned out he had lost a deal in Dallas and saw that his own money was in something of a precarious relation to his pocket. He quietly sold back his share of stock to me and removed the liquid asset to another account, with only his name on it. I was now sole owner of the water under a great variety of Texas churches, and indirectly proprietor of the Christian God of the New World.

I didn't care. I should say, it didn't upset me to discover that Murray had pulled out. He was someone who had visions but no tangible hold upon the real world. That was Murray's genius–to see it all in the abstract and live a vivid dream life in pursuit of things that never actually existed. Aren't all brokers great dreamers and idealists?

I went out onto the empty streets of North Zulch at around midnight. Me and the bats and the little noises in the grass. The houses were dark and humming from the air conditioners. Here and there a flicker of silvery light where some insomniac stayed up to watch the late show. But quiet everywhere, dark, deep country quiet. I sat down on the road in the darkness between two distant lamp poles and felt the sun's heat come up through my haunches. A mosquito hummed past my ear and went on to some open bedroom window.

Just then I heard a crisp little click on the pavement behind me. Not hard, tentative and withdrawn, as if a ghost had only one small shoe to walk on. Then another click, three at once, then five. Something with many small legs was walking sideways behind me. I had no sense of threat, though I felt blind sitting there. I didn't know. I couldn't see any shape. The moon was gone behind thick clouds. A whole sequence of clicks, some higher than others. A chorus line?

I stood up without a creak and breathed softly. There was no wind now. Something brushed close, almost against me. It was warm. It was high as my hip and the movement caused me to see a shape, a vague brownness in the blue of night. *A deer!* A herd of them passing around me. The bucks were ahead. I could make out the harder, lower thud of their hooves now. The lighter ones came behind, going around me. Not even bothering to sniff me. They mistook me for a doe or another buck. The little ones walked together, very slow and stately. One did slide against me, just barely, and dragged its hoof over my shoe.

I felt a deep, glorious thrill go through me. I could smell the grass in their breath, the bark mush, the scat under their tails. Large bellies and black eyes passed around me. I stood alone in the garden of Eden, all by myself in this Indian night world. A blind Adam, incapable of beholding the ecstasy of paradise. I could only smell, feel it. My brain couldn't really help me much. I was *there*, in the garden, but I couldn't see it. And there was Murray always seeing it and never getting there. Poor Murray, poor old make-believe Murray.

———○———

That morning, our last at the digging site, was a white summer scorcher. By eight the shadows had crawled up under the porches; some wild pecans had shed their leaves in the August torment. The bare yards looked almost like winter in the hot, dull rage of the dog days. Fall, winter and spring were all compressed into a kind of mini-season in late March and April, and then the killer days came to wipe out the world and leave nothing green before the first blue norther raged down from Oklahoma. It was already ninety by the time we got the winch motor turning. The last little section pipe got screwed on and a crowd showed up.

Word had spread that I would make or break today and the old farmers lined up against our orange fence with humorous looks and winks. Maurice came out of the gas station bathroom with shaving cream on his ears. He looked fresh and ready to work. He wore the same old greasy jeans with the towel jammed into his back pocket. Off we went again, revving up the motor and spinning the pipe down into the last mysterious levels of inner earth. She bit and staggered and spit up more red sand and pebbles, and Maurice cursed at the trash blowing around the pipe mouth. "Just more shit," he said to himself.

But a second sound began to roar under the moan and screech of the bit and the winch motor. A bass chorus began to join in and fill the air above us. It was a strange noise and I looked around for planes or a large truck. There was the crowd but it looked up with the same idea. Then the soot blowing down out of the pipe gurgled and spit up black splotches and then gray ones, and then long syrupy brown ones.

"OIL!" yelled one old skinny rancher with his hat held close to his chest. "It's oil!"

The crowd held out hands to catch a little of the sooty moisture tumbling down on us and someone rubbed the stuff on his fingers and cried, "Water! It's water!"

And it was water, brown, dark, hidden water. The well digger came alive as if it had reserves of power deep inside the earth to draw from. The motor sang at a high pitch and the pipe swung round more evenly, singing a steady high note.

"She's bitin' on water now," Maurice said low and breathless. "It'll come up any second now," and he had his hand on the big release lever.

"Here she is," a woman shouted and stood back with the rest of the back row, as a great thunder clap at the pipe mouth announced a huge dark funnel of water that rose, bent in the air, and came down in a slow, dissolving arc of rainfall over us. The water poured out in thicker knots, and then shot up in long, steady gouts of cold, icy white clear water. The water ran down the pipe and onto the truck, over us, over everything. The ground darkened and our feet and pants grew soggy standing there.

"Mister, you got yourself the best damn water in the county," a red-haired man said cupping the water in his hand and sipping it. "Best damn water in Texas," he said, laughing to his neighbors.

Maurice had run around to his tool box and brought out some valves and pipe joints and was busy pulling up spare pipe now. The water just came up of its own, running in great long swaths out of the pipe hole and coursing down the grass into the parched gutters.

The kids began dancing around in the cool moist grass, and two or three of the younger women let themselves get soused in the spray until their shirts clung to them. They were braless and their dark nipples shone through the thin cotton. No one seemed to mind as they danced around and gathered up the hems of their home-made skirts and let their white knees show. Their feet got muddy and smacked the soft wet grass as they hummed and did a little buck-and-wing. The old women smiled. Even the Reverend Hazard took off his glasses and rubbed his face with the cold water in his hands. "Praise God," he said, and the women dancing nodded and said "Amen."

It was good. It was like last night among the deer in the pure blackness of Eden. I felt giddy, as if I had downed a shot of whiskey. The earth swam around me. The dark sea of God had come up into air like a bolt of renewing consciousness. We were filled with grace. We were not strangers now. Everyone smiled at me and patted me on my back. I had brought miracles.

"What you aim to do with all this here good stuff," the feed store man asked me. I knew who he was because his shirt was sewn thick with patches of different feed brands. He had seeds on his pants and a smell of dry hay and horses.

"Sell it, I guess."

"Sell it to me, then. I'm buyin'," the man said, taking off his red cap and scratching the large freckles on his forehead. "Hell, I could use everything you got down there, and more."

"I got more, I got hundreds more wells like these."

"Is that a fact," he said, looking down and getting shrewd with me.

"That's a fact," I said.

"Well, now, ain't we got a talk comin' to us."

"Nah," I said. "I'm not anxious to sell it right now. I think it's time to give thanks and forget about who owns what."

><+>-0-<+><

It didn't interest anyone when we capped our well and put a nice secure valve and lock on it, ran a pad of concrete around it, and a fence, and made arrangements to shore up the church on its new footings. I put up a sign that read, "Warning. No trespassing. Water Well No. 1 - North Zulch Station." We did all that without a crowd. But the talk was thick all day in town, and the rest of the week. An article in the paper said a man found an underground aquifer in North Zulch where none had been expected. A few engineers from the state came out and estimated I had a goodly lake of fresh water down there.

I know, I said. And they asked me how I had come to dig just there. Certain facts gave me the answer, I said. I was not too talkative and they got the picture. It was worth a lot of money, all that good irrigation and town water, they said. Could hook up with the water grid from other stations and make a tidy profit if I would cooperate and do it the state's way. Hmm, I thought. Nah, no interest. I'll wait, I said. And off they went again in their white station wagons and yellow helmets.

The second well came in at Gause. Maurice began to think different of me. He didn't curse as much and he learned to trust my suggestions. We brought that well in under four days. I knew then all my water rights were good up and down central Texas. Every one of those churches stood over the clear pure waters of God. Murray had figured it out. I would give Murray an interest again, if he wanted it. He didn't. He was through with the business, he said. No telling what the consequences of meddling with secret underground water and religion would be.

⸺◦⸺

Fine. I would go it alone. I bought up a few more rights from some of the churches near Madison, and in Burleson county, and along the Brazos. There were black towns in the bottom land of Old Spanish Road and it was harder to buy up those rights. The preacher there, a man named James S. Riley, a Lutheran minister from Waco, told me the church couldn't sell its own birthright or its water rights. He was angry when he said it. Then he looked at me and said, "God is everything you see out here. He is the water and the air as much as anything in the sky. I believe that even if I am a Lutheran. If you love your soul, son, you'll give all those waters back to their spiritual guardians and renounce your deeds. These waters are God's pantry, his cookie jar for the children to come. He means to feed the souls unborn. We've wasted and squandered and ruined our great gift and these are his hidden promises. Give them back, and trust in the will of the lord."

I went home to Bryan and sat up in my study all one lonely night pondering this and other matters. Nothing could persuade me to give up the water. It was mine. I could make a fortune from it. I could be one of those chosen few who make real money and rise up over the drone of the working classes and find leisure, power, subtlety. Oh to be a water sultan, a grand Turk of the moisture abounding here, below human reach. I wanted to be the rain king, the last pharaoh of thunder and floods. I wanted so much to know the feeling of freedom from bills, drudging, blind struggle. What would it be like, I asked myself, to float through life? How would it be to meet only the beautiful people, wear fine clothes, eat the best food, and have no worry cross my mind ever again?

I took out my map and studied it carefully. All my churches stood on the blue veins of the planet, over the godly waters. There lay the cosmic milk, the breasts of the underworld from which these dry, bare, peeling chapels and brick temples took nurture.

That was it. The little churches were the meager heirs of an ancient tradition of water worship. The dawn people had come upon such springs under the great boughs of the trees and found their gods in the dew of the grass. The fountains bubbled and gave the word of God to the faithful; a healing spirit always touched and cured the staggering believer who asked for it. The mouth pursed and swallowed the divine seed from every cup; a bit of sky and stars shimmered in every sip of holy moisture. He who hath water hath the Lord.

We hadn't come far to build little chapel boxes on the dry land, and to hold in reserve this sweet cache of life serum for the faithful. Across the burning landscape of August were the dots of steeples and patches of clapboard, the tambourine thump and organ grunt of worship over the little vein of silver in the ground. Once, long ago, a vast sea lay upon the land and bore its presence into the crevices and faults and seeped down into the darkness below. Then the sands blew and the wind rolled the sage brush and carved the bleak mesas of our dry future. The ant made his kingdom here, and the iguana and armadillo wore iron coats to hold back the daggers of the sun. The Indians wandered the earth and like their ancestors, lay on their stomachs at sunrise and saw God's mist rise through holes in the ground. There lay the sunken sea, the sweet blood of the land.

My head was clearing. I felt a strange certainty sweep over me. The terrible nightmare of ignorance was lifting. Beneath our sturdy little prayer boxes lay the great watery underworld of grace. The preachers ranted and cried and clapped their bibles to their knees and spun their believers into trances. They hated as much as they loved. But they stood on the sacred fountain when they preached, and spoke from a pulpit of water and hidden love.

Who was I to take their God away? The man who stole God. Would that be my epitaph? Nah. These were Christians in name only. They were really pagans, druids, wizards, cave men in their undying love of water and the gods who lived there. All spirits stay in the waters, said an old woman on radio once. When we poison our rivers they go away and leave us to die.

In the soft undulations of the springs, in their long breaths, in their murmurs to the rocks and the hidden places, the gods lived and spoke to us. The waters were in us, the blood pumped round through all its conduits and chambers to tangle up with the earth's streams. We were all children of the water, God's water babies.

I burned up all the water rights in my study, one by one, with the phone ringing from Murray's office. He had changed his mind. He wanted in again. I wouldn't answer until the ashes were cold. Then I would tell him about my theory of America. How we were all cursed to be individuals driven apart by self interest and delusion, but we ate the same grain, drank the same water that was equally the *udan* of India and the Hittite *watar*, the Danish *vand* and Gothic *wato*, the Russian *voda*, the Swedish *vatten*, the Old High German *wazzar*, and the Old Frisian *wetir*, the Anglo Saxon *waeter*, the Latin *unda* and the Spanish *agua*. Our mouths sucked at the same divine breast and our lives were bound by the one great string of life. We beat with one heart in the breast of nature. I would tell him to renounce his past. I would ignore his protests, his long, thin cry of outrage on the phone. He would yell once more that on every dollar bill the free masons had left their coded invitation. Not "In God We Trust," but "I Turn God West."

West into the parched sands of money and despair. No, Murray, come unto me that I may pour the word of God into thee and lead thee under the oak and the cactus for fasting and purification. Lie down with me by the still waters, dear Murray, my broken broker, my thirsting lamb.

Part II:

City Ways

I've written books about American presidents, and one Republican president whom I've enjoyed writing about is Theodore Roosevelt, known as the Great Conservationist. In a message to Congress in December, 1907, he had this to say: "To waste, to destroy, our natural resources, to skin and exhaust the land instead of using it so as to increase its usefulness, will result in undermining in the days of our children the very prosperity which we ought by right to hand down to them amplified and developed."

– Paul F. Boller, Jr.

Olive Hershey

Remarks on the Environment in Texas

In **Wharton County** the calves are dropping like fat, furry cookies into the lush clover. Indian Paintbrushes stand so thick along the highway right-of-way that the whole world seems to be blushing. Cattle prices are steady, and the price of natural gas, plentiful in this southeastern Texas county, is nothing to fret about. It looks like another sweet spring in Texas.

Until you consider what's on its way here. Michael Shelton, a Houston lawyer and his company, SEM, plan on trucking tons of petroleum waste to this mostly rural county, creating a hundred-acre depository of hazardous waste, despite everything Harold McVey and his neighbors can do to prevent it. Looking anything like an Earth-Firster, Harold is a trim, well-tailored, incisive man in his seventies, who used to work for Texas Gulf Sulphur. Years ago that company extracted most of the sulphur from a huge salt dome near the town of Boling, in the eastern part of the county.

Because he knew the history and geology of the Boling dome, Harold was sufficiently alarmed to round up a cadre of ranchers and farmers and merchants who call themselves Concerned Citizens Against Pollution (CCAP). These folks look with a jaundiced eye on the plan to develop Wharton's sagging farm economy by burying petrochemical waste in the salt dome. Shelton's company bought part of the dome and applied for a license to the Texas Natural Resource Conservation Commission (TNRCC) to construct a multi-million dollar plant near Boling.

McVey points out that petrochemical waste has never been stored in a salt dome anywhere in the United States, although natural gas has been stockpiled in salt domes for years. Back in the thirties and forties oil companies explored the area, affecting the stability of formations underground.

This particular site has over twenty thousand holes in the caprock, McVey remarks, and that makes it unstable.

Shelton's company is doing business under the benign-sounding name of Secure Environmental Management.

Speaking of euphemisms, TNRCC is another prettier name for Train Wreck, which is how most of us refer to the monster bureaucracy spawned ten years ago when the Texas Legislature folded the health and water regulators all together to make public health, general safety and environmental matters even more muddled than they were already. All three commissioners, appointed by ex-Governor, now President George W. Bush, have direct ties to the chemical and petroleum industries.

Nothing startling there. The revolving door has been a fixture of the regulatory architecture for a long time. The surprising thing about this particular local outbreak of environmentalism is that the project, or something very much like it, has already failed to secure a license from the TNRCC. Under a similarly ambiguous title, United Resource Recovery, to which Mr. Shelton had leased the Boling site, applied for a license ten years ago, McVey says. Luckily for CCAP, Wharton County officials, being the wonderful, tight-fisted conservatives that they are, had managed to amass a budget surplus. Because of that surplus CCAP could hire a lawyer and a geologist who out-talked and out-fought the fancy lawyers on the other side.

Seven years after their defeat, the same people, the same company with their shiny new name, rode back into town armed with new legislation. In the next-to-last session of the state Legislature, they succeeded in pushing through the Legislature, with the help of Baker and Botts, one of Texas's most powerful law firms, a law enabling the TNRCC to consider watering down some of its requirements for permitting a hazardous waste disposal site.

Before the new legislation, purveyors of hazardous waste had to show an urgent public necessity existed for a Class I

Injection Well to be constructed. Now it's only necessary to show economic necessity, i.e. the need for refineries and oil companies to find a dump site closer than Louisiana or West Texas, where wastes are currently stored. Up until now, three dimensional seismographic pictures of the complete underground structure of a salt dome had to be taken by any company looking to bury hazardous wastes there. The TNRCC's proposed new rule would require 3-D on just the portion of the dome containing the waste, saving them a bundle.

You have to wonder why the same project under a new name can reapply repeatedly for a permit to install virtually the same dubious project. Harold McVey and his neighbors worry about their groundwater. They believe the old stricter rules offer more protection than the new more liberal ones. Three dimensional photographs of the entire dome, McVey says, would best determine whether the underlying structure could safely contain the hazardous waste. After all, oil companies commonly use 3-D seismic in Wharton County to tell them where and in what quantity the oil and gas is present. Shouldn't the TNRCC be asking dealers in hazardous wastes to use the same accepted procedure to determine if the underlying geology is safe?

The average citizen's only buffer against a million-dollar company purveying toxic and hazardous materials is the TNRCC, whose purpose is to mediate between the needs of business and those of the average Joe. One has to ask, finally, what kind of protection are we talking about? What kinds of decisions has the TNRCC made in favor of the little guy?

How far back do we go to answer that question? The last big decision in favor of Joe Citizen of which I am aware was to deny a license to bury nuclear power plant waste in Sierra Blanca, Texas. In 1999, after administrative law judges heard thousands of hours of testimony in hearings that went on for years, and after the state agency trying to build the dump spent five million dollars of taxpayers' money trying to secure a license to bury radioactive materials, the TNRCC turned them down. The two judges wrote in their recommendations that they weren't convinced the State's science was sound. President George W. Bush is fond of pointing to that decision as an example of how well environmental protection works in Texas.

Two big reasons the farmers, ranchers and merchants in Sierra Blanca won that battle were two very small groups led by Bill Addington and Linda Lynch. When these West Texans heard that the State of Texas was bent on bringing radioactive waste from all over the country into one of the largest, poorest counties in the state, they got organized. Educating themselves concerning the very complex technical aspects of nuclear materials, their half-lives, their destructive potential, Addington and Lynch got mad.

In the annals of amazing victories by environmental Davids against industrial Goliaths favored by politicians, the Sierra Blanca story has to rank as one of the most hopeful, a bright spot on the otherwise disastrous landscape created by pro-business environmental regulators in Texas.

But there's no need to get too elated about Sierra Blanca's narrow escape from rad-waste because some of the same pro-nuclear lobbyists are back in the Texas Legislature. They're pushing new laws to assist Waste Control Specialist, a private company based in Pasadena and owned by Harold Simmons, a generous supporter of President Bush. WCS wants to bury nuclear waste in Andrews County, West Texas. Under this scenario waste will come from many sources: including the U.S. Department of Energy.

This time the waste peddlers are positioned to win, though hawk-eyed local citizens in West Texas are well-informed. Their hackles are up, and they're getting help in the Legislature from folks like Bill Addington and Clark Lindly, who fought and defeated the Sierra Blanca dump three years ago. But for every voice of protest, there must be hundreds on the side of building the private facility. For every dollar raised to publicize the oncoming tide of Department of Energy waste along with unlimited waste from any state in the country, there must be thousands supporting the politicos and lobbying law firms which drafted the proposed legislation.

It's enough to make a person want to just lie down and cry 'uncle'.

But we can't, especially in Texas. Because of the way the rules are written by the TNRCC and our Legislature we must stop tending our garden and working our cattle and raising our children to pick up our swords once more, giving battle to the same people, the same companies over and over and over until, finally, someone writes some truly descent laws governing the licensing process. Texans desperately need laws that protect ordinary people who can't afford to move out of their neighborhoods once radioactive and chemical waste have fouled the land and air and water, ordinary people like Harold McVey who write letters and hammer at the doors of government and industry, year after year after year.

Pat Carr

Oil Camp

Corroded metal skeletons of the rigs tower above the company houses while jack rods pump backward, forward, grooving the arid dirt with an animal whine. Endless pipelines curve beside the roads, across ravines, until their rusting joints and elbows disappear into the foothills. The rainbow slick of a sump hole at the end of the row of tanks always holds the upturned stiffened legs of a cow or sheep or deer who slipped into the black pool and who has to be lassoed and pulled to the bank – always too late. Odors of oil and natural gas clamp like a lid over the minuscule town, cling to school sweaters and Big Chief tablets. The flavor of crude can be detected in Sunday apple pies and in radishes from the gardens. The alkaline water dries in concentric white rings at the base of the cottonwood tree by the bridge across Salt Creek, and when the last puddle vanishes, the ground splits into the dry, baked fissures of overdone cake.

Mark Busby

The Sacred Hoop

The Superconducting Supercollider was supposed to be built right through the middle of the place that holds my first memories, the old home place near Reagor Springs, just outside of Waxahachie and Ennis, Texas. My hometown of Ennis doesn't have many memorable places or history, it seems to me. Not like Waxahachie, anyway.

WALKS-A-HATCH-EE. Even the name sticks in the memory. I always heard when I was growing up that the town had been named by an Indian. The story was that on the very spot where the courthouse was built an Indian saw a newly born chick walking with part of the shell still stuck to its back. The Indian, the story went, in Tontospeak, pronounced: "Ugh, chickee walkeee before it hatchee," and thus the name was born. I liked that story and it made sense. Even the team mascot picked up on the story, for the Waxahachie Indians in green were dead solid antagonists of the Ennis Lions.

In high school, I was disappointed to learn that Chief Waxahachie's grave supposedly lay just outside of town, because it challenged my chickie story. Some of my football buddies and I drove out there one night to do what high school rivals do. We were going to dig up the old chief and hang the corpse from the courthouse before the big game. Somehow members of the Waxahachie team got wind of our plan, and they were waiting near the hill that was supposed to be the chief's burial mound. We drove by them real slow and yelled out names and gave them the finger, went down the road and stopped. They then drove by us real slow, yelled out names and gave us the finger, drove down the road and stopped, and we then drove by them real slow and yelled out names and gave them the finger, and you get the drift. It was pretty hot stuff, we thought, and we went back and told stories about how we'd had a rumble and beat the shit out of the Waxahachie boys. Waxahachie won the game 35-7.

Still later I learned that the mound we were going to ravage was just a small hill and that the town was named for a Waxahachie Creek that runs through town. There wasn't any Chief Waxahachie. The Tonkawas had named the creek for the cows that watered along there, and the Chamber of Commerce translated the name as "Cow Creek," but a friend of mine who researched the language says it actually translated to "Cow Chips." The Tonkawas were probably calling it Cow Shit Creek to remind them that the water was fouled there. Anyway, the flat land with gentle creeks appealed to cattle drivers, and the old Chisholm trail had passed right through the center of town. Years later down near Spindletop this side of Beaumont, I saw some real Indian mounds, squat hills with oblong crowns, nothing like the one here.

Waxahachie became a mecca for filmmakers in the early eighties when Texas was hot and the oil boom smoking. *Tender Mercies, Places in the Heart*, and *Trip to Bountiful* had been filmed there, partially because the state film commission was working hard to sell Texas as the third coast and partially because the old Victorian houses with broad front porches with swings and gingerbread found ready backdrops in Waxahachie. There are wonderful old houses with the wrought iron fences and a great square and court house. It's a Romanesque revival model, built of sandstone and gray rock with turrets – Texas Gothic circa 1895. It ought to have been one of those traffic circles, I think, instead of a square. It was always a rush to drive the traffic circle in Mexia, to watch for the openings between trucks, and to avoid the pickups that ignored the Yield signs. But Waxahachie has a square. It's slow cause you have to stop for red lights, but it gives me time to get another good look at the courthouse. It's changeless, the gray rock cut from underneath the black land that made Ellis County the biggest cotton producing county in the nation.

Just down the highway are Mansfield and Midlothian, Middle-of-the-ocean we'd say in that thirty seconds it took to drive through town on the way to Fort Worth, or we'd joke about how people only half-assed loathed the place. Later I learned that John Howard Griffin had lived near there after he published *Black Like Me*, but I didn't know any of that at the time.

It's easy to see why this area was selected for the Supercollider; it's as flat as a disk. The new road is as smooth as a piece of paper. They were to build their circle fifty-three miles around here, from Waxahachie arcing to Ennis and Midlothian and Maypearl and back, and it's all as level as a straight edge. And not too many people were to be put out either, just a few old farmers probably.

The SSC was to unlock the mysteries of the universe and be like Maxwell's demon, a self-generating dynamo that produces energy in perpetuity. We were to learn how to make energy from hydrogen and extract sunbeams from cucumbers. They were going to build an underground tunnel 150 feet below ground and use magnets to get quarks, those subatomic particles named by the one scientist with a sense of humor, going at the speed of light and have them collide with one another. That was supposed to produce something like the Big Bang, and then we would know more about the creation of the universe and about mass. It was to change everything we know about science.

Afterword

The SSC died out with a whimper of politics, the same way it came in. Now its skeleton lies moldering along that fifty mile circle where they'd bought up the land and forced people out of their homesteads. I haven't been back since I went out there and got a shotgun pointed at me and focused my consciousness considerably after a layover at DFW. Bought me a palmtop computer and a digital cell phone. Still fly through DFW fairly often. The last time I flew in we circled in from the southwest, right over where Chief Waxahachie's grave was supposed to be, and I had the faintest glimmer of travois, horses, and dogs moving south, and I thought long and hard about what we've done to the land.

William Hauptman

Wonder Pool

After World War II, my father came to Wichita Falls, Texas, with a degree in geology from the University of Nebraska and the intention of making some money in the oil business. The country near Wichita Falls had been the location of one of the great Texas oil discoveries, and there were still a lot of opportunities there.

When I was a child, my father sometimes took me into the field. We went to one of his leases, or wandered along the Red River between Wichita Falls and Electra and Burkburnett, which had been, in the early 1920s, the location of the Texas Wonder Pool. Once this had been a forest of wooden derricks, and there were still some old wooden derricks standing then, along with rotted wooden bullwheels, the rusted globes of storage tanks, and the cracked black dishes of slush pits. It had the feeling of a place where some great catastrophic event had taken place, like a battlefield.

The afternoon of my father's funeral, I accepted an invitation from one of his old geologist-friends and flew in his Cessna along the Red River and over the old Texas Wonder Pool. Seeing it from the air, I realized for the first time how big it was. Every wellhead, and there were hundreds of them, was surrounded by a circle of exposed red clay where nothing would ever grow again. It looked like the surface of Mars.

Of course the country around my hometown was never all that beautiful in the first place. And the oil companies are now more sensitive to the environment. But I also know that as hard as you try, you can't take power out of the earth without leaving a trace.

Tracy Daugherty

Amarillo

What's a Sonic Boom among friends?"
"Well, it's somewhat startling, but very solid assurance of protection. It's an unspoken pledge that the safety of you and your family is the prime concern of the day-to-day business of the U. S. Air Force."

This ad appeared in the Amarillo News-Globe in 1961, on the tenth anniversary of the siting of a Strategic Air Command base in the dusty Texas panhandle.

By 1975, Amarillo's Pantex Plant had become the final assembly spot for all nuclear weapons manufactured in the United States. At its peak, Pantex produced 1500 warheads a year – an average of at least four a day.

Many of the locals, hired to work on the bombing components, were fundamentalist Christians who relished the idea of the end of the world. It meant the establishment of the Kingdom of Heaven on Earth.

I'm passing through today after visiting family in Oklahoma City. A sign on Route 66, just across the Texas state line, says, "Rattlesnakes. Exit Now." From Clinton on west, billboards have been screaming, in blocky red letters, "The Big Texas Steak Ranch. 72-ounce steak. You'll be glad you waited."

In Amarillo, stoplights sway over wide, empty streets, clicking green to yellow to red to green to yellow to red. There is no traffic. Tumbleweeds. Caliche. TV satellite dishes. A cracked Dairy Queen sign says, "Amarillo. We Like Where We Are."

This part of the world – my big, gusty backyard where the bomb came to sit one day – has always crushed me with loneliness. These days, post-Cold War, it seems more desolate than ever.

For the casual traveler, with nothing personal at stake (but who would come here casually?), Fodor's sums the place up well: "As the last echoes of the Civil War cannon died, and men turned their thoughts westward," says the famous travel guide, " they aimed at nearly every corner of the map except the Texas panhandle. That way . . . lay only madness: the madness of the incessant howling wind (summer and winter); the madness of the unbroken prairie where a man might wander aimlessly until the blistering sun of the windblown dust finally felled him; the madness of icy snow, borne horizontally on the wings of a roaring gale. This was the Texas panhandle."

This was the Okies' escape route.

And in the early nineties, this was the mad track of Timothy McVeigh. Obsessively, he drove to and fro between Kingman, Arizona, where his friends the Fortiers lived, and Herington, Kansas, to visit his pal Terry Nichols.

Here, in this hardscrabble basin, hour after pill-fueled hour, McVeigh envisioned the bombing of the Murrah building. Is it coincidence, I wonder, staring now at cactus, weeds, alkali streaks in the sand, that this parched, dead-flat stretch of the road is also the beginning of the militarized West?

In the thirties, when the Okies sputtered through here, none of what McVeigh would later observe had yet been built: nuclear weapons plants, test sites, Air Force bases.

"Restricted." "Federal Property." "Keep Out."

Initially, the interstate highways that erased 66 from the map were developed mainly for the military. Dwight D. Eisenhower had returned from fighting in Germany highly impressed by the strategic value of Hitler's Autobahn.

In the High Plains, and in much of the West, the roads were designed to be emergency runways for fighter planes. A B-52 bomber from Oklahoma City's Tinker Air Force Base could land on a long, straight line of I-40, refuel, resupply, and be back in the air in a snap, defending America from the terrible Soviet threat.

Throughout the Cold War, most people who lived in

"hot" towns didn't mind the government's presence. Uncle Sam's pockets were deep. But to a migrant, unmoored, lacking direction – to an Army veteran, no less, who'd finally been rejected by the Army, as McVeigh had been – it was easy to believe that the U. S. government was a kind of occupying force, waging war on American soil.

Most farmers, victims of federal price-fixing, would agree.

Most ranchers, pinched by federal land regulators, young Ivy Leaguers in big, climate-controlled rooms back east, would agree.

And Native Americans? Workers in the fields? Well. They weren't talking. If they were, no one was listening.

Midland, Texas, my hometown, lies directly south of Amarillo, south of 66. As a child, watching Roy Orbison on my daddy's brand new Crosley, I wasn't aware of Pantex. I didn't feel the nuclear weight pressing on me from the desiccated plains up north.

But of course the bomb had changed my home.

I know now that most television and computer technology has developed through military research over the years. Satellite tracking. High frequency radio waves.

Though I wasn't aware of it in 1959, the big-eyed box in my parents' living room was a poor civilian stepchild of an angry warrior father.

All the while, as Roy wailed, "Ooby Dooby," the bomb was sitting just up the road, biding its time. In the hands of fervent Christians. In swales of blowing dust.

Today, with the collapse of Cold War money (the bomb in hibernation, its messes strewn recklessly across the desert, from Amarillo to Hanford), it's impossible to overstate the sense of betrayal one feels in many parts of the American West.

For example, in Hudspeth County, in far West Texas, the ag-based economy has been severely depressed since the late seventies. The area only averages seven to nine inches of rainfall, yearly.

In the early eighties, the federal government proposed turning the county into a nuclear waste dump, storing ninety per cent of the nation's radioactive waste there. The locals were not informed of this decision, and only caught on, much later, from newspapers.

Meanwhile, Merco Joint Venture Inc., a private sewage company, purchased a 128,000 acre ranch north of the proposed nuclear site, and began disposing sludge there – nearly 225 tons of it a day from industrial sewage plants back east. Merco told the citizens of Hudspeth County it was providing fertilizer for overgrazed rangelands.

At the edge of the Nevada Test Site, an organization called American Peace Test provides advisory bulletins for protesters: "The Nevada Test Site is a highly radioactive place with many hot spots, dumps, and storage areas . . . [T]here is little that can be done to protect your body from beta and gamma rays which are unseen . . . Cover your face when walking in the wind. Do not eat food dropped on the ground. Don't use bare, dirty hands for eating . . . Depending on [where you are] you will have to deal with ammunition strafing, falling bombs, unexploded bombs on the ground, maneuvering around targets, and stealth bomber[s] . . . [S]ecurity forces are well-armed and quite capable of shooting if they feel threatened."

Really.

Is it any wonder that many Westerners today, like Southerners after the Civil War, feel they live in a conquered nation?

Is it any wonder that a furious farmer and a disillusioned vet would plan to bomb a government building?

The wonder is, we didn't see it coming.

The miracle is, it doesn't happen every week.

From a speech of Harry Tracy Daugherty, my left-leaning grandfather (1935): "We must stop the destruction of true democracy and the substitution of government by force."

Timothy McVeigh (1993): "the violations of the Constitution by those power-hungry storm troopers of the federal government [must] not succeed again."

Michael Fortier (1997): "If you don't consider what happened in Oklahoma, Tim was a good person."

From the outskirts of Amarillo, a right turn takes me back to Oklahoma city, on I-40. Cornstalks scritch in the fields. High clouds, banked against the blue.

I recall a local television commercial when I was a kid in these parts. A pretty young woman smiled at the camera and said. "If you don't have an oil well, get one!"

See what I mean about betrayal?

Charlotte Waley

Dallas and the World

I was six when I arrived in Dallas, Texas, in 1932. My father was among the few fortunate employees with the Oil Well Supply Company who were invited to relocate during the Great Depression. We left Oil City, Pennsylvania, in early spring, drove south in caravan with other Oil Wellers on miserable rutty roads, and found a house in an area that is now known as the "M Streets." A few months later we were introduced to chiggers, Johnson grass, crepe myrtles, cicadas, and oppressive, mind-numbing heat. There was, of course, no air conditioning.

Nor was there much of anything in the way of scenic beauty around the flat lands of North Dallas. But there were jobs. Businesses grew, air conditioning made living in the torrid climate bearable, and the downtown skyline began to evolve. Mobil's flying red horse graced the tallest building and soon became the centerpiece of the city. Automobiles proliferated. Fossil fuel was good. The term *global warming* was not in the lexicon. After all, the oil industry helped to build Dallas and other great cities in the Southwest.

Seventy years later, soaring steel and glass buildings surround and dwarf Pegasus. Man-made lakes and parks, museums and music centers, manicured suburban lawns and gardens enhance the quality of Dallas living. Still, in 2001, downtown Dallas remains the scenic center of this sprawling city. While other places have their mountains and lakes and canyons and beaches, Dallas has its splendid sculptured architecture – covered many days now with a heavy curtain of murky gray smog.

As the result of its spectacular growth, the city has environmental problems, polluted air from automobile and power plant emissions being among the most serious. The popularity in recent years of large, gas-needy SUVs has exacerbated the problem. Ozone-alert days have become more frequent. Our local, state, and national political leaders have done little to encourage industries and citizens to curb heat-trapping smokestack and tailpipe gasses.

Courage is needed in the White House and Congress to prioritize the development of alternative forms of energy, cleaner industrial plants, and smaller, fuel-efficient automobiles. Opponents of such measures have protested that the economy will suffer, but the lead was taken out of gasoline some years ago and the economy survived. Now it is time to attack the other air pollutants – soots, sulfur and carbon dioxides, nitrogen oxides, and mercury, to name a few.

With all the angst in our new-millennium lives, we certainly do not need to contend with eye-burning air and contaminated water, not to mentioned floods and other catastrophes resulting from holes in the ozonosphere that melt polar ice caps. Our grandchildren should be given a chance to live in a secure and smog-free world.

To be fair, some industries have taken the initiative on their own. In the DFW Metroplex, TXU is trying a new technology, a "smog-eater" to reduce nitrogen oxide emissions at its Lake Ray Hubbard power plant. And our old enemy Japan has already produced "green cars" with gasoline/electric engines that conserve fuel and reduce emissions dramatically. Such responsible efforts toward improving world environments are significant steps in clearing up our polluted air, but more must be done. (Whatever happened to research on nuclear fusion?)

Now seventy-six, I will probably not be around to see our leaders take much action to curb the poisons that cause global warming. But I am optimistic enough to think that the cleansing will eventually happen, given the dedication and determination of a farsighted few. The people of the world must be spared a future climate as hot and dirty as Dallas in the summertime.

William Harrison

Why I Live in Arkansas

When I grew up in Dallas it was an overgrown provincial town of 300,000 residents with six high schools, a minor league baseball team, and streetcars. Everything else, as I recall, was either a movie house, a park, a cafeteria or a church.

The air was so clean that out in Oak Cliff we sat on our porches on the hottest summer nights looking at the stars – hoping, for luck, that one would fall – and learning the constellations. My father's barber shop on Bishop Avenue smelled of bay rum and talcum. No family owned two cars. There was no television. Everybody took care of his lawn with those little bladed push mowers. My first job at age eleven was on the truck with Mr. Creel delivering ice. The alleys behind our house were rutted mud tracks.

When I went away to College in Fort Worth I took my date, JoAnne, to the prom out at the Lake Worth Casino and as she stepped out of my dad's old coupe into a mudhole she looked down at her silver slipper, sighed, and remarked, "I wish they'd pave the whole damn world."

They did.

Dallas became a metroplex crisscrossed with concrete freeways. They even paved the alleys. Thousands of cars speed along to superstores of windowless air-conditioned skyscrapers where the occupants watch television or surf the internet. Out at the new major league ball park the fans watch replays or read their statistics off a jumbo screen slightly smaller than the suburb of Oak Cliff.

Over my old city there is an inverted bowl of brown smog. At night the fumes from a million cars hold in the neon glow of the city, so there are not stars – or precious few.

Concrete plazas dot the countryside around Dallas: isolated clusters of immense office buildings with glass that mirrors other buildings nearby. The plazas have a surreal emptiness: cars parked out of sight underground, no pedestrians or even sidewalks, lawns inserted by machines and watered automatically, a lifeless and oddly lonely landscape.

Recently I spent a number of weeks in Austin, a city I remembered as smaller, gentler, and perhaps more habitable. But it's another Dallas. So is Houston and San Antonio. Midland, Lubbock, and a half dozen other Texas cities seem ambitious to copy the mold.

Texas has for a long time encouraged its cities toward an unrestricted growth that feeds on nature like a cancer. Its pollution laws are unenforced. Its water resources protections are virtually non-existent. As its oil, chemical, banking and insurance industries have thrived its environmental record has become a disaster, and now it is boldly exporting its brand of raw boosterism to Washington. We are all being invited to share the Texas catastrophe.

I've loved Texas and I've drawn much of my craft as a writer from its people and character, so that's why its descent into the future hurts so much. Beyond the urban sprawl there lies a fearsome, Blade Runner decay: nature not only ignored, but forgotten.

Marshall Terry

Lead (Pb)

We were up at dawn drinking coffee. I stood looking out at the metallic sun in the gray sky as my wife read the paper. I never got much beyond checking my stocks and looking at the sports – in retirement you're entitled – but she always read the morning news. Our great metropolitan daily planted winsome little feature stories about "real people" in its pages weighted with all the dreadful news like remnants of another age.

"Dear Lord, listen to this," she said.

I poured another cup and listened to her read a feature on a fellow we knew, old man by the name of Arthur Tillis who lived not far from us over by – well, pretty nearly right beside – the central expressway. Arthur had a sticker on his bumper that said *I Hate Central Expressway*. He had a fair-sized strip of land between his little old house and the service road. On it he had one of the biggest plot gardens – tomatoes, corn, okra, beans, squash, all that stuff – in the neighborhood, and there were plenty of little gardens around here. Most of us had lived here a pretty long time. Put it this way, there weren't many children in the neighborhood, at least before the behemoths began to be built. Anyway, what she read was that right at Arthur's was one point where the city had just now gauged the lead content in the soil and found it was one of the heaviest counts in the city, almost as high as the site of the old smelter plant on the outskirts of the city.

"Isn't that awful?" Grace said.

I nodded. It was terrible. Then I put on my Les Miserables sweatshirt and got my maddog stick and went out on my morning walk.

The concrete was cracked in the sidewalk in my block. In

front of some houses there was no sidewalk. That went back in a labyrinth of time to when the houseowner had that option. The old trees along the parkway were gnarled and unpruned. I had asked the city to prune the two in front of our house but they hadn't quite gotten to it yet. I turned the corner and started west on Milton. All these little streets were named for mighty poets. Old Arthur Tillis lived on Shelley cater-corner to Central. I would go east to Cambridge, which bisected the poet streets, then back west to Arthur's and on back southwards, home. It was a good two miles, about right for a senior citizen like me with a body like a peanut set on rods.

I went by Henry Bascom's. He had a nicely done garden built up into a square with telephone poles in his yard. He put some sort of chemical in his soil so it was nearly white, and grew huge cabbages and kale and asparagus and stuff like that. I never knew why anyone would go to so much trouble to grow cabbages. Along farther old Bill DeBussy sprayed every darn little bug that came on his plants. He sprayed so much he hardly got a tomato last year. I have a rackety little garden out back on a small plot behind the peeling garage. If I put in about fourteen tomato plants we get enough for two for salads through the summer. Early Girls and Better Boys do real well in the clayey soil if you mulch it good.

Turning on Bryon I saw old Fritz McGee and Hap Henry coming towards me on the other side. They walk about as slow as you can go but pretty good for two jokers at least five years over eighty. They wear old floppy hats and talk a lot as they walk. God knows how many hours it takes them to make their round. They are always out walking when I come out and always still walking when I'm on my last leg home. Hap always smiles and waves, the short, fat one, but Fritz, the tall one, never does. He always looks like he's mad as hell about something.

They both have canes and I see Fritz pointing his at one of these new brick monster houses they began to build along in here and sell for half a million dollars before the economy went kaput. This one is about four stories high and is all orange-ish brick. It really does look like some monster staring down at the little one-story clapboard houses still on each side of it. Fritz shakes his stick at it like it's some obscenity. Old Hap just smiles and waves to me, then stops and carefully picks up the newspaper lying on the cracked walk of the little house next door and with a pretty good arm tosses it farther up into the yard.

Old Hap and Fritz are like two watchmen of the neighborhood. Fritz is right about how ugly the huge new expensive houses are. Most only have two little round windows on the sides, like for machine guns. Most are standing empty now. I walk by the one Fritz pointed at and remember it's on the lot – these things take up the whole entire lot so there's no room for trees or a yard or garden – that my now deceased friend Bob Shaeffer owned. Bob bought his house and lot there thirty years ago, I know, for $14,000. Now it's some empty monstrosity priced at $500,000 they can't get a penny for. Whose idea of progress is that, I'd like to know?

Down the street Miss Phoebe Hoagland was standing in her driveway by her nice brick house with her hands on her hips. Miss Phoebe is pushing ninety. She takes care of her sister Miss Esther, who has been infirm for many years. Miss Phoebe and Miss Esther were both celebrated schoolteachers and from one of the original families in the city. Next to her house is another behemoth, this one all dark red brick but built by the same blockhead that built the orange-ish one and of the same exact design.

"They have cracked my driveway," she said. "The builders. I have had a concrete man here who says it will cost three hundred dollars to repair this. Their cement mixers and other heavy trucks used my driveway, with no regard for the fact that it was my property, and then –" she paused to look up at the sky, white-haired, with her bright blue eyes, like a fierce small bird, searching for the word that would say it right "– blithely drove away without offering restitution."

It was pretty well cracked up, all right.

"Well," said Miss Phoebe, squaring her shoulders and standing straight as she could with her humped back but well under five feet tall, looking me square in the eye, "if they think they can get away with this, they are wrong. They do not realize they are dealing with a Hoagland!"

I said I hoped she could get restitution and asked after Miss Esther. Miss Phoebe said she was not at all well and was in fact going blind, which was a blessing for Miss Esther since she would not have to see how these cavalier construction people who erected this incredibly ugly house now by them had cracked the concrete in their driveway.

The sun was burning off the grayness of the morning now. The sky was getting milky blue, and the sun had an orange edge to it. It was pretty even though you knew it was pollution.

I turned off Cambridge and walked west on Homer. All the other poet streets were English. I thought it was funny this one street should be epic Greek. I wondered why Miss Phoebe had allowed them to break the scheme that way. She must have been right here and teaching her famous literature course in the school back when they were naming streets. Or maybe she had done it, maybe she had wanted them all Greek. I walked back towards home on Homer, hearing the roar of all the traffic on the expressway before I got to the corner, to old Arthur Tillis' little house and big plot of garden.

Old Arthur was still a large man, real handsome, now with silvery hair. He was out in the garden with his wife, working away. He looked like he was happy as a lark. I was kind of embarrassed to stop, wondering if they'd read the story mentioning them and the lead in the soil and their garden. Then I thought I always stopped if Arthur was out in the garden, and I better stop now. Of course they had seen the story.

His wife was a small woman with a bandanna on her head. She was working a row of carrots with a watering can, looking serene. Arthur was checking on the cloth bindings that held his tomato plants to the stakes. He still used the old wooden stakes and soft cloths and not the wound wire supports I had gone to. It was early July, a few days past the Fourth, and Arthur's first crop of tomatoes had come ripe. There were about forty plants of them, all gorgeous, from the big Ponderosas and Beefsteaks to his Early Girls and little red and orange cherries. Also he had rows of corn now forming and beginning to tassel and dark green rows of bush beans and an orderly pattern of a lot of other stuff.

"Tomatoes look good, Arthur," I said.

He picked up a grocery sack from his old redwood table and tools and garden things.

"Here, take some," he said.

I did not know whether to laugh or turn red. Then I saw he was perfectly serious and was looking at me real intensely. I had known Arthur Tillis for forty years. He was one of the first ones to buy a lot and build his house around in here, years before they built the expressway right by his house. He'd been an engineer. He was one of the smartest guys around.

I said sure. He picked about half a dozen of the nicest-looking darn tomatoes I ever saw and put them in the sack for me. I thanked Arthur and his wife profusely, and headed home. The sun got hotter as I walked the last leg south to my street. It was going to be a bitch. I held what Arthur had given me in my free arm up against my chest as I went along the sidewalk and thought I could feel the tomatoes ticking in the sack.

Cathy Tensing

Texans, Come On Out and Look at Yourselves!

You don't really see Texas when you live here; you have to go away for the summer like we do each year, to live in an old southern French village, in order to come back with a critic in the back of your eyes, that urgent "catch" right above your heart that keeps you looking for what you thought was here.

In France, we live simply, in an old rhythm, buying fresh fruit and vegetables from the old-fashioned supermarket in the next town, a chicken from the local poultry farmer, eggs from my friend down the road. And olives in the Saturday market, the green ones sprinkled with fresh garlic, the black wrinkled ones from Nyons. And wine from our friendly wine seller, who knows the vintners by name and has stories about every bottle you choose.

Summer goes by slowly, and I am always outside with village friends, visiting their gardens, sipping tea under the old plane trees as we while away an endless afternoon. No stores, no gas stations, nothing but the old hills as the Romans knew them, and the scatter of stone houses, and a few newer bungalows.

After arriving at the George Bush International Airport, driving toward Houston, I found myself looking with a little bit of fascination at all the shopping centers.

My "jet-lagged" brain got back to glancing at billboards, and speed reading road-side signs as we sped by . . . until at some point I realized how many times I had seen "We Buy Ugly Houses." That's when I started to feel irritated, overwhelmed by all the stuff being hawked, all the stores urging me to hurry in and buy, buy, buy. I didn't want anything, but here was a whole corridor of signs on both sides urging me to consume more than I wanted, more than I could ever need.

There was smog over Houston. It was Sunday with not much traffic; and soon we were on Hwy 290 heading for

Austin. Finally, when we took the turn onto Route 6 the sides
of the highway were Texas "treegreen" again and puffy clouds
kept me staring and marveling most of the way to College
Station, where the car dealerships and malls once again lined
the highway.

Then it was the quick-stop gas stations with arctic air
smelling like bug spray, and shelves sagging with every kind of
sugary, salty, starchy snack for making you gain weight, and
all those national franchise restaurants that pump the smell of
grease into the air to lure you in. We stopped at the new
Albertson's supermarket, newly built across the street from
Kroger's and next door to an HEB Pantry, and we were the
only customers. While I shopped, I breathed an odorless air of
sanctity among the vegetables, glittering from the mist just
sprayed down on them. The meat counters were packed with
perfectly symmetrical trays of steak, chicken breasts, turkey
wings, and nothing smelled like food. The rest of the enormous
room was lined with canned and boxed goods, legions of them,
as if sheer multiplicity could possibly mean variety. It doesn't.
In fact, there is no variety in modern, post-industrial
America. We shop for spaghetti among forty brands, but the
differences are nil. We merely compare unit prices, and go on.

We were coming back to my husband's job at the university and our "big, old wooden house" near downtown Bryan,
the home we bought 30 years ago. As I got back to watching
TV, I was fascinated by all the new commercials. It looked to
me like advertisers had finally caught on to the fact that
nobody smart watches the ads. The first one I noticed was for
paper plates – telling us women to use them so we won't have
to stand doing dishes after dinner. The plate was shown to be
strong enough to carry food on it to the next room. It took a
bit of dissembling to have an "American family" playing a
board game instead of being glued to the tube. And on the second take there was mom with her husband and children. What
happened to the French tradition of sitting through a four or
five course dinner? Of hearing those dishes washed and warm

with delicately sliced cuisse de canard arriving after the cream crab-stuffed avocado of the first course; with plenty of time in between to digest each other's ideas and taste the cool Chablis. Didn't Americans ever hear on TV that you don't have to rush back after the commercial?

It might feel good to take a walk after dinner, say hello to your neighbors (if you nod to them through the closed air-conditioned houses), hear the night call of a bird or two, or just walk down a dark tree-lined road where no cars speed by. But there's no immediate safe place, no road that hasn't been built up. No space that's "just there," not being used for anybody's profit . . . or fenced for a dog whose bark scares you when you come near. In America, if you want to go anywhere, first you have to get in your car.

I spent my summer in Europe sitting in sidewalk cafes in Lucca, Italy along with handsome, well-dressed people with dark eyes. They hadn't given up the pleasure of sitting down, drinking espresso, eating homemade apple tarts, and talking to each other. There were women who had just bought themselves bouquets of flowers from the open air market, with fresh cut wedges of cheese wrapped in white paper and a baguette that already had the end of it broken off for a taste of that hot, soft flavor. Friends greeted each other with hugs and a kiss on each cheek as the waiter held his tray of glasses of red wine or Perrier poured over ice and a lemon slice.

But at Albertson's, I noted a retired couple sitting sheepishly at the little table next to the deli buffet – they were eating sandwiches for supper, drinking coffee out of styrofoam cups. They weren't talking, not even looking up, for fear someone might censure their behavior. They looked and felt out of place doing what any normal Mexican or European would do any time they felt – at an outdoor restaurant or cafe. Here, it seemed strange and awkward to be eating in public, with all those faces staring down at them as they pushed carts along.

In Paris there were things to do that you didn't have to pay for. The Mayor had put sand in the plaza in front of the 'city hall' and volleyball courts to draw people out. All the time there was a huge movie screen for watching soccer games and along the Seine River beach chairs were arranged with a constant water mist that cooled sun worshippers and in the evening brought friends out of their hot apartments. Plus hundreds of cafes where people of all ages meet and talk and "live".

Oh Texans, you have lost so much – what are you getting for nothing; with your highways crisscrossing over Houston and Austin. Your expensive cars, your sealed houses, all your land bought up and fenced. You have consented to too many shopping centers; too much merchandise that you have been told will make your reality. I see it in your eyes. You don't know wild horses, deer, antelope herds, buffalo, wolves, rattle snakes, clean rivers; the wilderness that used to make you feel you were alive, where you could rope it in. You have abandoned your small towns where you knew the teachers of your children, where your politicians knew who they were hurting when they let the big corporations buy you out and then pollute – take and not give anything back to you.

I see it in your women, some fat and couched in their living rooms; some nervously thin, covered in make-up and expensive slacks; each trying to find some place to sit down and feel like she belongs, where she could be part of a village culture again; a place to be with other women who think their lives are fun, who have hope and live in solid, stone houses where spirits and muses still inspire them, where the TV antenna blew down last winter and nobody bothered to get up on the roof and get it working again.

You park your car and run in for a quick buy, then out again, racing off somewhere. Home to that TV program you have to watch each Monday? Or to pick up the kids, or to get back to work. What have you settled for? Why have you let life keep you so busy? I guess with your "small town" heart, you didn't realize what was being done to you – or who was taking over your country, your rivers, your air. Your freedom.

Come on out and take a look at yourselves.

Part III:

Natural Ways

The natural environment is the cradle that holds our stories. Every story has to settle in a place and put down roots if it is to have any mood or gravity. My stories are anchored in their places, and whatever they lack in character and plot, the vegetation is always right and the landscape is drawn to scale. I don't consider the landscape part of my fiction, but part of the fact on which my fiction is created, as solidly important as any of the historical events I dramatize, or any scientific truths. It isn't to be played with. I don't put rattail cactus where it doesn't grow.

– Elizabeth Crook

Reginald Gibbons

Refuge

Beyond the suburbs, armed
bulldozers crush the libraries
of wildflowers, religions of
the butterflies are suppressed,
hamburgers force their way
even into the forests that
never before had been
cut, there they impose
administrative districts of
the bewildered star fields
of bright trillium that
constellate the ground . . .
But farther in, farther for
a little while yet,
the sister saplings still do
not tell their myth, they whisper
warnings to each other, blue-
feathered heads abruptly
look up from their reading
of acorns, tall leafed
beings breathe out
benevolence and each grass
stem no different from
innumerable others happens
to move in the soft sweet
air in such a way as
to be fully itself, singular,
it sways gracefully,
alone now, alone, unafraid,
the center of the world.

Reginald Gibbons

Manifesto Discovered Under
a Fringed Gentian

On the foundational dark silence
        of the forest
and without need
        of worshippers
temples grow with architraves
        of reaching branches that rest
across columns
        of air and live sap, decorated with living
friezes, moving scenes in the lives
        of leaves.

Therefore proclaim with the voice
        of moss
an inviolable protectorate
        of wrens
and another, other-continental,
        of toucans.

Confer most-favored status, for the trade in air,
on all remaining mountainsides.
                        Array in fractal
regiments the sands and grasses
to guard the weary shorelines.
                        Administer
last twilights to rivers which, if they are still
believers, may be reborn.
                        Patrol
the poisoned deltas and lake bottoms and sea depths
with bottom-vigilant finned inspectors.

                    Sort
through the gull-shadowed garbage that has buried
jaguar nations
        of the badger, frog and trout,
and tie the old tin cans
        of rusting gold mines
to the long tails
        of CEOs
                Lock down the wilding
International Monetary Fund and transpatriotic corporations
        on diets
        of clean water and animal crackers.

Herewith we abolish the practice
        of board-feet and clear-cuts
Hereby we establish the free speech
        of bristlecone branches and song sparrows,
Hereafter we will publish only anthologies
        of wildflowers,
We hereupon enjoin all poisoning of owls,
        of wolves, winged fish eaters, orchids and heads.

Till the new republic
        of the rain in common
can celebrate the green tally
        of a one hundred per cent
turnout
        of the blades
        of the grass of the fields,
        of the leaves
        of the trees of the forest.

Ron Tyler

Smog on the Horizon

The sudden appearance of Ruidosa and the road leading to Marfa jarred me back to the reality of my situation – I was still miles from my destination and, for all practical purposes, lost, my sketchy map being of no use at all for this leg of the journey. I turned toward the northwest, hardly anticipating the spectacular mountains of this vast ranchland – mountains that forced me to slow down out of awe as well as danger. As I wound up and down the mountains and through the sandy valleys, I drove over cliff roads that could not have accommodated two vehicles had I encountered anyone. Chinati Peak to the south of the road dominated the view, and as I climbed down from the mountains I was forced several times to pause and take in the beauty of this virgin country.

Finally, as I roared over a small hill, with dust plume in tow, I hit the promised pavement and blew past a waving border patrolman, no doubt lying in wait for the smugglers I had wondered about. Realizing that I, in my rented car with out-of-state license plates, probably looked more like a smuggler than anyone he had arrested this year, I pulled to the side of the road as he approached. I described my day's activities in enough detail that he motioned me on, and I was not too distressed to notice that he took down my license number. Once again on the road, I tried to calculate the time now required for the drive to El Paso. I had given up hope, because I was still miles from Marfa. Only then could I accurately judge the time it would take for the rest of the trip.

But, wait! As I topped the next hill, there was Marfa, laid out in the valley before me just as clearly as it was in the many nineteenth-century glass-plate photographic negatives that I had seen of the city, which was founded as a way-station on the

Southern Pacific in 1882. The jagged peaks of the Davis Mountains and Mount Livermore, more than 8,000 feet high, were clearly visible as a backdrop to the city. I could not be more than five minutes away. I might make it yet.

As I whisked confidently along, dialing the radio in hopes of picking up the local radio station, I passed a road sign indicating that Marfa was more than thirty miles away. I had been badly fooled for the second time. I later learned that the Davis Mountains were another twenty miles beyond the city. On this clear day I had a stunning vision of at least fifty and perhaps one hundred miles of the Big Bend country, laid out before me in convincing detail. No wonder the McDonald Observatory is nestled into those mountains – to take advantage of some of the clearest skies in the country.

Because of the pollution in the skies of the Big Bend today, I fear that view is no longer possible.

Jerry Bradley

How the Big Thicket Got Smaller

First they got the panther,
then the bear.
The pigeon flocks
went with the wild timber,
and oil,
plain crude,
drove out the dens of all sorts.
Roads did in the plants.
There are still some trees
though now they stand
in a slaughter house of saws,
and iron rigs
black as poachers' kettles,
bob like chained birds
drinking from the earth.
A life ordered and lubed,
hard as blacktop,
a tax-supported legacy
for the mechanical world
where the ivorybill
once tapped its confused code
and big cats screamed
like frightened women in the dark.

Michelle Pulich

Wild Texas

This big state is and has always been as intriguing as a big bright oil painting, especially in terms of what is natural and wild. Sea anemones on Port Aransas jetty rocks come in oranges and pinks. And when the tide reveals the colors, you can watch them open and close, small silvery minnows winnowing by on breaths of salt water. Not far inland, patches of remnant coastal prairie wave little bluestem and Indiangrass in the breeze, like the stubble and wispy hairs of the earth, masking black, raw soil beneath.

There are so many things I love – little details about wild Texas, that I want to tell you about – like how on April afternoons in Live Oak County, tarantulas, big as a gorilla's hand, sun on country roads. Like how delicious agarita and huisache blossoms exude their odors in brilliant yellow when spring hits that brush country. Like how coyotes sing regularly amidst a chorus of pauraques and nighthawks under a clear patch of the Milky Way, just outside of George West at Gussetville, next to the old Spanish cemetery.

Wilderness may not be the first thing that comes to people's minds about Texas, but elements of the wild abound – and definitely pre-date us here. We owe honor to the wild that remains.

There is also what once was here to remember. There is a rare plant, Walker's manioc, a subshrub in the same family as cassava, that I was fortunate enough to see in the field in Starr County in the lower Rio Grande Valley of Texas. As far as is officially known, there are only twelve populations of it left in the wild.

If the cells that compose Walker's manioc no longer exist on Earth, I won't be able to soak up the leaves' snowflakes shapes with my eyes. Crouching on haunches over the rough ivory sandy loam, no one will be able to examine the bendable, rubbery nature of the smooth gray stems. I also fear the deaths of my ailing friends. Maybe this is a part of why I absorb the ones I love as much as I can when we're together because I know our time is finite. At night, unasked, unbidden, I have dreamt of the Starr County rare plants. In sleep, I made a magenta Star cactus tea and deciphered the swollen cactus buttons, as others have read tea leaves. In sleep, I walked through forests of Walker's manioc, staring at the deeply incised leaves like they are cloud shapes, perhaps holding secret answers. On the chance that every second really does count and that a moment echoes forever, then every life would count and the other biotic parts that in their varied assemblages, have been and will be, matter as much as your next breath. In this world with change and love as the only certainties, the ache of nonexistence will throb.

Lisa Sandlin

We Are Here Together

I grew up in the bayou country of East Texas near chemical plants and methane flares, where a footstep pressed on water, where frogs and crickets sang like a tide. In the spring and summer the singers held a continuous note – or new ones joined in as others dropped out to breathe. That's how I thought they worked their seamless chorus, singing and breathing and singing and breathing. Not an accurate image, but, for a kid, a joyful one: Frogs breathed by singing.

Summers we visited Gatesville, Texas: mesquite and pale dust, rocks and blistering sun. Out there I learned I wasn't dangerous: not one of the horny toads I caught shot blood from its eyeballs at me. An uncle warned they did that. I scooped them up anyway as they hunkered down disguising their horns and spots and round brown bellies as dirt and scrub. What a marvel. Thrilling to pet with their spininess, their dry skin, rough and soft. Breathing. Much better company than baby dolls or hoola-hoops, better than pink plastic high heels, horned toads were alive. Real. They were creatures like me. We were here together. Certainly I was messing with their sun-bathing, ant-eating day; they scrabbled. They also lay in my hand, cocked spiny heads, blinked, considered. I plunked them in a slat crate and talked to them under the mesquites' skinny shade. Come evening, I turned them out again.

The Texas Horned Lizard and the mountain short horned lizard are now protected by law. A child today couldn't converse with near as many as I rounded up in the '50's. Many scientists report that the frog malformations studied in Minnesota and elsewhere result from fungicides, pesticides, chemicals in environment, contaminants in the water. "Frogs are storytellers of water," said a biologist I know, "and since they live there, they suffer quicker than we do." It's in the tadpole stage they're most likely to suffer deforming changes. Bill Moyers' documentary on the chemical industry noted the awful parallel: "Children are the most vulnerable" to pollutants. Hard not to reflect how in our family, my grandparents, parents, and my generation were born healthy, but the children following us have suffered spina bifida, cleft palate, kidney failures.

Another biologist told me we can have a win/lose – she said we keep doing what we're doing and win our short-term comfort, but eventually we lose by killing off water, habitats, species, and our descendants. Or a win/win – we preserve our environment, even if we have to sacrifice, and we save living land, water, animals, and ourselves.

"I hope our whole nation can come together and start treating each other with dignity and respect," Laura Bush said the other day on TV. I was cheered to hear her say it. Mrs. Bush was talking about people. But lawmakers concerned with future energy sources should also consider the integrity of the earth and the lives of their grandchildren as worthy recipients of Mrs. Bush's' good wish for mutual "Dignity and respect."

Diversity? The natural world is the source of that concept. Tolerance? How can we apply that word solely to human beings? Surely, we're smart enough to crave the amazing differences and purposes each species owns and the richness that brings us. Maybe we can even remember that we're often more generous people when we're up against it – a storm, an energy crisis, a national sacrifice – and we have to work with each other for an end we want. It's not just a kid's point of view to say: We *are* all creatures. We are here together.

Pattiann Rogers

Being Accomplished

Balancing on her haunches, the mouse can accomplish
Certain things with her hands. She can pull the hull
From a barley seed in paperlike pieces the size of threads.
She can turn and turn a crumb to create smaller motes
The size of her mouth. She can burrow in sand and grasp
One single crystal grain in both of her hands.
A quarter of a dried pea can fill her palm.

She can hold the earless, eyeless head
Of her furless baby and push it to her teat.
The hollow of its mouth must feel like the invisible
Confluence sucking continually deep inside a pink flower.

And the mouse is almost compelled
To see everything. Her hand, held up against the night sky,
Can scarcely hide Venus or Polaris
Or even a corner of the crescent moon.
It can cover only a fraction of the blue moth's wing.
Its shadow could never mar or blot enough of the evening
To matter.

Imagine the mouse with her spider-sized hands
Holding to a branch of dead hawthorn in the middle
Of the winter field tonight. Picture the night pressing in
Around those hands, forced, simply by their presence,
To fit its great black bulk exactly around every hair
And every pinlike nail, forced to outline perfectly
Every needle-thin bone without crushing one, to carry
Its immensity right up to the precise boundary of flesh
But no farther. Think how the heavy weight of infinity,
Expanding outward in all directions forever, is forced,

Nevertheless, to mold itself right here and now
To every peculiarity of those appendages.

And even the mind, capable of engulfing
The night sky, capable of enclosing infinity,
Capable of surrounding itself inside any contemplation,
Has been obliged, for this moment, to accommodate the least
Grasp of that mouse, the dot of her knuckle,
        the accomplishment
Of her slightest intent.

Pattiann Rogers

A Passing

Coyotes passed through the field at the back
of the house last night – coyotes, from midnight
till dawn, hunting, foraging, a mad scavenging,
scaring up pocket gophers, white-breasted mice,
jacktails, voles, the least shrew, catching
a bite at a time.

They were a band, screeching, yodeling,
a multi-toned pack. Such yipping and yapping
and jaw clapping, yelping and painful howling,
they *had* to be skinny, worn, used-up,
a tribe of bedraggled uncles and cousins
on the skids, torn, patched, frenzied
mothers, daughters, furtive pups
and, slinking on the edges, an outcast
cow dog or two.

From the way they sounded they must have smelled
like rotted toadstool mash and cow blood
curdled together.
All through the night they ranged and howled,
haranguing, scattering through the bindweed and wild
madder, drawing together again, following
old trails over hillocks, leaving their scat
at junctions, lifting their legs on split
rocks and witch grass. Through rough-stemmed
and panicled flowers, they nipped
and nosed, their ragged tails dragging
in the camphorweed and nettle dust.

They passed through, all of them, like threads
across a frame, piercing and pulling, twining
and woofing, the warp and the weft. Off-key,
suffering, a racket of abominables
with few prospects, they made it – entering
on one side, departing on the other.
They passed clear through and they vanished
with the morning, alive.

Steven G. Kellman

Aversions to Pastoral

My house, an echo chamber in which five thousand books clamor for attention and contention, sits on a plot of 1.9 acres. The glare of the Texas sun spares me the need to don reading glasses, and as often as I can I squat outside the house and entwine myself, clear-eyed, in the plots of authors who sit continents and centuries away. Outside these paragraphs, butterflies flit, and stickweed creeps. Oak wilt, and roses climb. Spotted, a doe darts off. Fire ants extend their claim. Indifferent to the pages I read and write, leaves of Johnson grass cloak a snake.

Scorning what he called "the idiocy of the rural life," Karl Marx sat daily at the desk in London deep within the British Museum. Is urban sprawl the postscript to *Das Kapital?* Is the legacy of revolution no place left to breathe? Marxism triumphs not in the gulag but the shopping mall, our monuments to the power of macadam to pave over Eden, which never was paradise, unless paradise is not perfection but rather loamy life. Nor does Adam Smith's fantasy of an invisible hand that kneads private greed into public good contain a green thumb. Capitalist utopians promise plenty and deliver blight.

Texas still contains wide open spaces, some not yet tainted by toxic wastes. On a track I made in our backyard, I still find room to run away from traffic. Everyone needs an unpaved acre or two to think, and, sitting in my garden, near a highway that serenades me with the whoosh of SUVs, I yet can think about the still truths that lurk beyond momentary gain, for the moment.

Wendy Barker

Trash

"TRASH," he said, as we walked
the line between our almost-country
properties. Again I pointed, trees, shrubs
whose names I didn't know, but "trash,"
he said again. Whatever wasn't oak.

That neighbor knew three kinds of trees:
live, pin, and Spanish oak. The rest should go.
Developers all clear that way. Chainsaws
take out everything not oak, and not just
cedars, the invasive, weedy junipers

that throttle any plant nearby. It all
goes: the buckeye – magenta blooming
in the spring – and husiache – yellow.
Mexican persimmon's bark's a subtle
complex painting, nuanced as a Turner,

and that's gone too, its little leaves
preceding tiny purple fruits that draw
the birds that nest on into June,
buntings and the warblers above
the grasses: gramas and the blue stems.

Council members said they'd leave the trees
when clearing for a new, expanded city hall.
But like that neighbor years ago, they meant
the oaks. And now they've called a meeting:
oak wilt has hit the neighborhood, and

oaks are what we're left with. Too much
construction, trimming of the trees, their
wounds not treated. The virus travels
through the maze of roots, connecting.
And once a tree's infected, it's just trash.

Wendy Barker

Full

Light splotches on the bed,
mesmerizing the morning.
Why rise from this dazzle?

But outside the kitchen door,
the first time in years, flickering
in the pittosporum's froth, a dozen

dozen Monarch butterflies ignite
the green, their white freckled patches
shifting, rapid as a blink, and gone.

Not so the evening primroses
that open as the light is leaving
and remain even as the moon lifts

from the trees, even as you sit
steady above your book, until
you rise, and bring me your hands.

Laura Furman

The Woods

For a long time after the Comanches had been driven out, not many people lived in the woods or the limestone hills. There was no easy way to get their from the rest of the city until, forty years ago, a man who owned some of the hills persuaded the city to build a bridge across the river. The chosen site was at a dam, and the cliffs on both banks were blasted through to make a place for the new road. When the bridge was completed, a new road wound upward to the hills and valleys, and people moved out, building houses for themselves. New trees grew and maidenhair ferns covered the scars on the cliffs. Just below the dam was a turbulent pool where kayakers practiced for bigger whitewater. Down river, at the opposite bank, fly fishermen cast where smoother water flowed. Between the banks was an isle covered with scrub trees and rocks. Often people fished or drank beer there, throwing sticks into the river for their dogs to retrieve. When the floodgates were opened, the isle disappeared, all but the tops of the trees.

The neighborhood in the woods was zoned so that the woods would be preserved, though the zoning and preservation were compromised by shifting times – trees did come down, houses did go up above the treeline, small and large infringements – still the woods were protected. Overprotected, some said, for what was the use of protecting the cedar trees that took the water and light from the oaks and sycamores? Beyond the city, on the ranches, the cedars were cut on a regular basis but here they were allowed to take over and become a hazard. If the neighborhood ever caught fire, the cedars would fuel the flames beyond control.

In addition to the plague of cedars there were the deer.

Trapped between the river and the new highway to the west, herds of deer wandered through the woods, browsing,

standing still and blink-eyed, close to the people who soon got
used to them. Between the cedars and the deer, the neighborhood in the woods had grown barren. Get ambition, grab a
shovel, ram it into the ground with all your might, and you hit
bonecrushing, spinerattling limestone. There was no soil, just
a deceptive layer of rotted leaves. Even native plants that deer
were supposed to despise disappeared. The serious gardeners
fenced their lots, acres to a half-acre, brought in truckloads of
topsoil, threw in their old vegetables and fruit composted to
richness, and made a garden and a lawn to that, driving along,
you saw the brown of the deer territory, the green claimed by
the people. Most people lived with the devastation or tried to
establish shrubs and plants the deer would ignore. Deer families crossed the roads in their perambulations, and the people
who lived in the woods soon learned that when a fawn crossed,
the doe wouldn't be far behind. The does were cautious and the
fawns were not, so you had to wait and wait to be sure you
could drive on. Somewhere, watching over the enterprise,
might be a buck who crossed even more slowly, bearing the
weight of his antlers. Strangers to the road or impatient residents hit deer, sometimes the careful drivers did as well. The
deer flung themselves suddenly out of the woods onto the path
of the car or stood waiting to be killed. The worst was when one
limped off, for the deer would die if lame and unable to browse.

When they first moved to the neighborhood, Josh liked
to play in the woods with his friends. Carlotta would sit on the
deck, reading and listening for their voices. Once, the boys came
running back to the house, excited. They'd found an arrow! It
was a deer arrow, a rigid steel shaft with razor blades embedded
at the tip. Carlotta reached for it but Josh wouldn't let go and
for a terrible second it seemed that his hand was grasping the
razor blades. He finally relinquished it and she promised to keep
it in a safe place. She talked to other mothers, and they too wor-
ried about letting the children loose, for other children had
found cruel arrows. It was illegal to hunt there, but that didn't
seem to stop anyone. A few women also confirmed what Carlotta
thought she'd mistaken: the report of guns.

Betsy Colquitt

Trees and Progress

This is their last spring, these trees,
and untutored by hollow-eyed houses
with jacks as new foundations,
these soon to lie earth level
bud, their branches gold
in promise this warm afternoon.

Borers, ants might have worded them
their doom. Progress too sends messages
but not such as trees can hear, heed,
and this afternoon, they stand
apriled and foolish, budding, leafing,
brilliant in their confidence of June.

Pat Little Dog

Speaking of Priorities

It happened sometime in the spring – after long hours of watching the rain drizzle in through a wide gaping hole in the ceiling of my closet apartment, catching whiffs of the raw sewage backing up outside its bathroom window, mulling over the sixth or seventh notice from the landlord about the water to be cut off again for pipe repairs – when the realization came to me that I wasn't going to be able to live in the city anymore. It wasn't just the discomfort of living in one of the cheapest apartment complexes downtown. The truth was that even at cheapest I was having a hard time affording it. There had been a string of disappointments – a failed publishing deal, a manuscript too personal and political for market-savvy editors of the times – all of us housing impaireds have some kind of similar story. Even with the invitations I did receive to contribute to various literary and regional publications, I couldn't write enough to live more than a subsistence life anywhere in my native state with its rents and utility bills escalating at such a terrific pace, the downtown apartment leases going up every six months' term by as much as sixty dollars while the rate of pay for free-lance writing of all types remained remarkably unchanged at the same low rates as fifteen years ago. My roommate Mac helped with the bills, but recently his health had gone bad – his lungs filled up, his heartbeat became urgent, his knees swelled up three times their normal size with water – bad for anyone, but economic disaster for an on-call waiter-bartender. And of course I was paranoid that the ill health which was hampering our economic efforts was at least in part related to the apartment being in the basement and the air conditioning ducts blowing into the tiny room vile air filtered as it was through the sediments from the rooms above us and the drain-offs from the parking lot pavement which extended up and out from the ceiling drip.

Later that year, after the first big cold spell hit, I read with great interest about the Austin writer Lars Eighner – how he had been found in his tent on the Greenbelt with his room-mate and his dog. One of our mutual editors, the soft-hearted Lou Dubose, then of *The Texas Observer*, had taken in the lit-tle household when the temperatures had dipped into the twenties. Then in the following days, fundraisers were held to get the three tent dogs into apartment living again. There was a photograph of Lars accompanying one of the benefit reports, his hand in the air as if he were about to be sworn in at some kind of a hearing. In fact the cut-line quoted his vow: that he would turn his writing skills to the (said-to-be) more lucrative mystery market – a promise, I suppose, made as part of a larg-er resolution to look to his own rent in the future, to come inside, sleep indoors, and to eschew the personal insightful street-revelatory taoistic autobiographic he had been known for, which had a particularly iffy market in an age addicted to the prescribed and programmed. He had thought he would be dead from his various ailments, pleurisy and such, before his royalties dwindled so he wouldn't have to deal anymore with the exasperating need of the living to feed, shelter and clothe themselves. As I read the pitiful little paragraph, I wondered what his expectations might have been when he chose the writ-ers' life – maybe different from my own ideas culled from a lit-erature produced out of seedy hotels and writer's garrets such as those of Henry Miller, Dostoyevsky, Francois Villon, the prison musing of Eldridge Cleaver, Henry Thoreau; then the rediscovered hardy women writers – Meridel LeSeuer coming home to an attic apartment full of hungry children – Zora Neale Hurston asleep in an unmarked pauper's grave for so many years before Alice Walker found her again.

"Why do we write? Because we must," Jan Reid once intoned, quoting the old writer's saw to the little group of us who had been stood up at a bookstore super-signing after the

publicity of the event had failed to go out – and so we had
drunk our free coffee, celebrating our own selves and our editor Billy Bob Hill who had even paid us money for our Texas
stories. Jan was the only one present working for a living wage
as a writer – a *Texas Monthly* man. One was a professor. One
was married to a professor. I myself had begun to live in a tent
– like Lars. Mine was not a dome tent like his was, but a larger four-room my mother had given me for my birthday.

There was another difference. Rather than setting up in
the Austin Greenbelt, I had bought my own land – a ten-acre
plot of the boggy woods of Caldwell County – the investment of
the last good chunk of money that had come my way – some
land to call my own being my dream of a lifetime. I had bought
farm land like my grandpaw once cultivated. He hadn't been
a land owner – instead an East Texas sharecropper. And my
father had not followed in his footsteps, but had left the fields
for the War in the early Forties, and like the song says, how
can you put them back on the farm after they've seen Par-ee?
Seventy per cent of Texas residents were rural before that
War; only five years later the percentage was down to forty,
with so many taking to the more entertaining, salary-spending
city life. But even though I got no heritage of land, I was given
a rich store of childhood memory coming from my stays with
my grand-parents between my father's military tours of duty.
I still treasure the world view my grandpaw gave me between
the ears of a mule he let me ride while he made the rows; the
wandering paths through the adjacent woods; the dark glimmer of the water at the bottom of the well; roads of hot summer powdery dirt so delightful between the toes; a fireplace in
the winter to heat the front room and roast sweet potatoes in
the coals; sugar cane to spit off the porch after the sugar had
been sucked from the soft splinters. . . the kind of stuff the
country ex-patriates still sing about in south city bars.

Long before my apartment ceiling had begun to drip
down on me, I had tried to extract lessons from my father's
choice of lifestyles. He had been what was known as a bird

colonel when he retired from the Air Force. He had dropped
down on the floor of his huge computer terminal in the midst
of his personnel – almost a thousand – in an underground
bunker honeycombed with huge data machines – a heart attack
at the age of forty-nine. I try to imagine the level of pollution
that he worked in – the totally enclosed building with the recir-
culated air, the adrenaline and stress and sickness of the military workers, the fumes of the processing fluids and maintenance chemicals. Then, too, he was overweight and under-
exercised, a desk man used to Officers' Club fare: lobster din-
ners, imported wines, cigars and brandy, the after-dinner
choice of diplomats; enlisted men opened and closed his doors.

His face had become red and puffy and his waistline dis-
tended before I left home. And yet he was given a full disabil-
ity, I guess because the military felt that the debilitation was
necessary to the work he performed. He didn't actually die
until he was seventy-eight, but he was constantly ill, in and out
of the military hospitals. They did everything to him – opened
up his heart, changed out his arteries a couple of times,
pumped out his lungs. He had taken so many drugs in his last
years that his glands had swollen to the size of grapefruits each
side of his neck, which was the poisoning he finally died of –
called lymphoma. His body had become so backed up with toxins that he was in constant pain, could hardly move. Some
people would say that modern medicine had managed to
extend his life to the age of seventy-eight, but I believe that his
life had been cut short and made painful by bad environmental principles. After all, his own father had lived to be a
healthy ninety-two.

Environmental ignorance is never becoming, whoever
bears its insignia. In the last days before I left the city streets
to move out into the woods, I saw a string of grade school children holding hands while waiting to enter a downtown theater
for their field trip – an afternoon matinee. They were all just
as pale and wan as spaghetti noodles, quiet and dazed with
Ritalin, eyelids blue and red from neon overheads and com-

puter screen lights. Their appearance struck me hard, seeing
them like hot-house plants with totally underdeveloped meta-
bolic functions. I myself already had two grandchildren of my
own – what legacy would I give them?

I've been living in my country home for three years now.
the original little clearing that Mac and I first made has been
enlarged to accommodate a garden and a duck pen. The tent
became a little shed made from construction materials recycled off of Austin's building sites at clean-up time, then dou-
bled its size to a two-room house. And company has come to
live here with me, including my daughter Brook and my three-
year-old grandson.

This is not to say that my land is a place of isolated pristine beauty. I got what I could get with what I had. The scars
of an earlier oil drilling operation are still very evident. Five
rectangular ponds have been abandoned on one end of the
acreage, plus a huge rusting tank and many feet of rusted
pipes and black hosing. I haven't tested the soil or the water to
see what kind of a mess was left behind when the oil men gave
it up, but tales of their early contamination have sprung up
from the neighbors' gossip all around me like the clear water
that used to fountain before it was sucked and salinated, then
left for dead. A mineral lease still falls like a long shadow
across the whole plot. But like the developer said who sold me
my parcel, that is the state of virtually all the land in Texas.
The local water company has assured me that the underground water table, only fifteen feet below the surface, is totally polluted from the run-off happening across the county of
chemical fertilizers and herbicides, and that the springs the
little town just down the road from me was named for no
longer exist, disappeared – or were covered over – when the
first oil drilling in the vicinity occurred, some of the earliest oil
prospected in the country.

And all around me the cedars have begun to die. First,
brown patches appear in the green foliage; finally the whole
tree turns dry brown. Maybe it's the polluted ground water.

Or maybe it's air pollution. Springtime has come again, the time of the year when napalm and paraquat are used from Texas south through Mexico, Central America, Columbia. Farmers south of the Rio Grande continually try to send up the message that coca and marijuana plants cannot be isolated from all the others – what poisons one will ultimately poison all. The smoke from all the burning fields gets thick this time of year, blowing as it does through the wide wind corridor that follows the spine of our continents, south to north, connecting all of us living along it. It doesn't take the long career and cultivated nose of a nature scout to recognize the ill health contained in this early spring fume.

But there is an even closer possible source of pollution directed specifically to the cedar itself. I found a folder on a feed store counter a few months ago advertising the virtues of a certain herbicide to help those interested in cedar and mesquite eradication – a few drops'll do ya'! – either dripped onto the soil or sprayed. The correct way of spraying such lethal liquids is one of the classes taught by the local agricultural extension agent, but this pamphlet assured me that I could buy and use as much of this stuff as I wanted without being licensed. I think quite a bit of this particular poison has been sold and used in my neighborhood in the three years of my occupation, what with so much land being massively and quickly cleared for the migration out of the city into the countryside again with people like me trying to get back to their rural roots, and so many of us accompanied in our re-entry with city ignorances and prejudices we have a hard time recognizing for what they are. This includes the belief that cedar is a "bad" tree. Cedar gives us fever – that's what the weatherman tells us – allergies – that's what the drug companies say. But I know better now. When I lived in the city, my nose and eyes ran, and my lungs filled with fluid. But when I moved to the country and learned to sit among the cedar trees, and even to purify my cabin of its mildew and molds by swinging pots of smoking cedar branches through the rooms, I realized that the

cedar tree was just as sacred as it ever was, and certainly not
the culprit when it came to my allergy symptoms. In fact, the
trees seemed to be my relief – after I removed myself from the
traffic corridor of IH35 and the core of the downtown congestion I had once chosen to live in for many years because of its
grooviness, without regard for its contamination. In gratitude
for the cedars' curative powers, I have begun to think of this
habitat as a refuge for cedars, which is why their die-off concerns me. Once there were cedars in Lebanon – now there is
only desert – why does this tree eradication plan continue to
roll around the globe?

In fact, I see my little ten-acre plot as a general refuge for
other creatures fleeing encroaching developments. I wanted to
make it official when I first came – to declare several acres as
natural wildlife habitat with the agricultural exemption which
is allowed under Texas law for such reserved wildlands. But
the county tax appraiser responded with scorn and rejection,
even though I presented a full wildlife plan including pending
renovation, identification of naturally growing herbs and dye
plants, and the maintenance of dewberry and agarita thickets
for natural feed. He said my land was nothing more than poor
cedar and mesquite thicket like all the other land around it
and nothing special about it to preserve.

I've proceeded with a little refuge anyway even though I
pay full taxes now on my land. I had to pay a roll-back tax
when I was turned down on my habitat plan consisting of the
difference between agricultural exemption and market value
on ten acres for the five years preceding my purchase of the
land – a substantial sum for me. "Just buy a few steers," I've
been advised by more than one, including the chief tax
appraiser, "or lease your land for grazing." The cattle lease is
what kept this land under exemption before me even during
the oil exploration days. "Drop the pitch about wildlife and
apply again in three more years." But today the woods remain
a tangle around me, my writing studio a round table and a
chair in the larger thicket.

Yesterday when I came out at sunset to herd my ducks out of their open yard and into their pen to lock them up for the night, my dog was making a ruckus, pacing the fenceline where the duck yard ended and my garden began. When I saw what he was barking at, I froze, staring eye to eye with a very large bird, its wings half extended, standing in my garlic patch. His dark head was turned directly to me, and his eyes were so set back into his skull that they appeared to be no more than black holes above a larger darkness where a beak should be, the late afternoon angle of light shadowing his head in such a way as to flatten the features. I thought at first he was an owl or worse, an ominous winged messenger, some border chupacabra come to speak to us of doom. I began to move slowly off the path and walk along the garden fence to see how he would react. At first he didn't move his head at all. I quickly saw that his beak was much longer than an owl's. In fact I now recognized him as a large hawk! I thought to myself once I got past the first impression – the awful apparition. And so even though he is moving slow, and strange, he's probably come after my ducks – even though they are Indian runners with necks almost as long as swans and wings as wide as his. Perhaps he's sizing them up, sidling up to them rather than doing the fast plummet down which I have witnessed and taken to be characteristic hawk behavior, that hawk dive that can bring a long dangling snake or writhing rabbit up and out of the bushes and into thin air in seconds.

He moved his head slowly when I passed him, but kept the rest of his body still, remaining in an awkward half-sitting position among the green sprouts with his wings unequally splayed. By the time I had moved almost directly behind him, his head was also turned almost one hundred and eighty degrees around; then it snapped back and for a moment he was fully turned away from me, his head sitting high on his shoulders. My dog had stopped in front of him on the other side of the fence with his nose to the bird and one paw in the air, pointing – or saluting – not barking anymore. Now that he

had told me to come out and I had come, he was waiting to see
what I was going to do.

I walked into the duck yard first and herded the birds to
the safety of their pen before turning my full attention to the
bird in my garden. Then I got my bamboo stick from the side
of the tool shed and slowly began to walk toward him, waving
it between us. At first he turned sideways and cocked his head
and looked at me with one of his sunken eyes. Then he walked
– more a kind of a stride than a hop – through the garlic and
over into the cilantro, then heaved himself up with only partially functioning wings to the top fence railing. When I kept
walking toward him, he half fluttered, half climbed up to the
top of the duck pen. I banged the tin with the stick. He tried
to hunker down on the opposite end and ignore me, but I
wouldn't let up. The ducks began to holler as he hunkered and
I swatted. Finally he swooped down into the open yard and
strode behind the tool shed dragging his large wings behind
him. My dog followed him at a distance. My three cats had
appeared, clumped up together at the clearing's edge to also
watch. When he stopped, we all stopped. He started out again
and got to the edge of the woods, where he swung himself onto
a low tree branch and hunkered down again. We all turned
then and left him alone.

Inside the house Mac was washing dishes and Brook was
making coffee. When I told them about the bird, Mac went out
on the porch to look. "Is that him?"

I came out to see where he was pointing. Sure enough,
the hawk had taken a position in direct sightline to the porch.
Mac brought a chair out and took up watch on the bird which
blended into the mesquite and hackberry trunks so well that
you had to know where to look in order to see him.

"The messenger," Mac said – because he comes from people who still think like that about such creatures.

That got us all to talking. Taking the part of the skeptic
now (after my initial reaction), I said that I didn't think he was
any kind of messenger at all, but had simply been hit up on the

highway and had limped into our clearing. But Mac didn't think he looked like an accident victim, more like he had been poisoned by something, the way his feathers were falling out and the stiff way he moved. Brook believed that he might have actually come because he knew, somehow, that this was a refuge. That seemed even more far-fetched to me at first than the idea that he had been sent. But when I thought again about where I had first found him – not in the open yard with my plump birds but wallowing down in the middle of my garlic patch. Maybe I had read the situation wrong from the beginning – he had never had an interest in ducks – it was the healing herbs that he was after.

There was still a little stew left over from supper. I put it on a plate and took it out to the woods. He had moved a little farther into the trees and tall grass. I placed the plate in a high fork of branches and left him on his own.

That evening we made a fire in the clearing and sat around it as we have done many evenings before. We talked about the sick hawk some more, as well as some of these other stories – the history of cedar, my father, the colonel who took to growing vegetables again in the springtime after he retired. Finally we trailed off into serious fire gazing, the stars peeking out above us, drawing out our deepest feelings, desires and dreams as starlit nights and home-built flames have always done for people like us.

It's obvious, I suppose, that my priorities and values are not in line with what our current culture is all about. I want more trees and less pollution. I live in a house wholly built from what city construction crews have dumped on the ground. I use very little electricity, even in the sweltering summer, when I spend most of my days either digging in my garden or writing in the shade of this grove. When I had the choice, I invested in woods and land instead of a computer. I still enjoy the use of a manual typewriter. I do not desire my own Web page. Maybe if I did, I would make more money peddling my name or my work, but a garden full of real green stuff, backed up with several acres of nopales, wild plum dewberry and other edibles, gives me a new-found sense of security. I'm reaching out now, at least, to a greener future, putting strong medicine in the air.

Bryan Woolley

In Jeff Davis County

I: The Davis Mountains

Some fences, some roads, some houses the Apaches
would rather die than live in, if they had not already
died. The grass is shorter, but Espejo would recognize
the stones. And Nicholas, and Quanah, and Victorio.

II: The Hawk

I killed you on this mountain twenty years ago.
My bullet pierced the soft feathers of your breast,
your sunlit wings, extended to embrace the sky,
crumpled. You fell like a meteorite. I ran
to you, exulting in boy-manliness, then backed away
from the angry stare of your yellow dying eye.
I pumped slug after slug into you there among the rocks,
afraid. I still hear their muffled thud, the rifle's
echo from the cliff. You soar above me, watching.

III: The Dead Cow

You lay in shade of a live oak tree at the foot
of Casket Mountain, ragged hide pitched like a Bedouin's
tent, dry and empty. Coyotes spread your bones around
you long ago, to bleach, to be no more a part of you.
The child inquired about your eyes, and how you feel.
Buzzards ate your eyes, and you do not hurt.
You are falling into the ground to be grass,
and there will be another cow.

IV: Mt. Locke

On Mt. Locke you gaze at stars and shoot laser beams
at the moon. British scientist in cowboy clothes,
you explain it. The telescope entertains us with how
small we are. Harvard's machines in Cook Canyon say
cattle will die of what is happening on the sun.
We are grass and rain, Victoria knew,
and Espejo, and Nicholas, and Quanah.

*I wrote this poem after a visit back home to my mountains about 25
years ago. Since then, the City of El Paso has purchased several
ranches there for the ground water under them. In a few years, the
water will be pumped out of the ground and piped to that city, and
the fragile aquifers in that semi-arid land will be sucked dry.
Meanwhile, two coal-burning electric power plants just beyond the
Rio Grande in Mexico are polluting the sky over the Big Bend and
the Davis Mountains, obscuring their magnificent vistas and spoil-
ing the air. The pristine beauty of the land that the Indian chief
Espejo and Nicholas and Quanah knew, and that I knew in my
youth, is dying. Since there aren't many votes out there, no politician seems to care. But we are still grass and rain, whether we
acknowledge it or not.*

Clay Reynolds

People and Wolves

A ndrew **Big Hands** pushed his horse into a biting north wind and, at the same time, tugged the brim of his hat down. He wore a scarf of heavy wool snugly wrapped around his neck and face so only his eyes peered out into the tiny, stinging pellets of ice that were beginning to fall from the lowering clouds. He had been out most of the day, and soon it would be too dark to see, time to return. His horse balked when they were struck by a gust of wind at the top of a short rise of ground, but Andrew urged him on, down into the depression and along a line of hackberries and cottonwoods which kept a tenacious hold in the loamy soil. He couldn't go back until it was full dark. He was searching for wolf tracks.

This was dirty work, and he knew that his boss, Mr. McSpadden, had given it to him on purpose. He gave all the dirty jobs to Andrew on purpose. McSpadden came into the bunkhouse that morning while everyone else was preparing to enjoy a day off playing dominos and watching the weather deteriorate. He was a big man who wore an expensive leather coat with a fur collar and a fine, high crowned hat. When he came in trailing an icy blast from outside, he grinned and asked Andrew to stand up. Then he put his hand on Andrew's shoulder and told everyone that Andrew could track animals better than anybody he had ever seen, so he had a special job for him. Andrew knew that was a lot of bullshit. He would have known even without the knowing grins that flitted around the room while McSpadden talked. He knew that the whole scene was just McSpadden's way of reminding him of who he was, what he was.

Andrew Big Hands was one of McSpadden's cowboys, but he was also an Indian. Andrew wasn't sure, though, exact-

ly what sort of Indian he was. He had been left tied to a cradleboard at the door of a reservation orphanage when he was still
a baby. He still had the cradleboard, but it told him very little
about his lineage. Both Kiowa and Comanche – or what was
left of them – lived in the region, but the beadwork on the
cradleboard could have come from either, or neither. For all
he knew, he might be Navajo or Arapaho, Cheyenne or
Cherokee or Apache. All that mattered to McSpadden and
most white men, though, was that he had nut-brown skin and
dark eyes and sleek black hair. McSpadden always called him
by his first name, as if he was an idiot or a child. Andrew told
himself he didn't care, that he was used to the attitude
McSpadden showed him, and he never complained about
receiving every dirty job that came along, accepting it all with
a nod and a grin. There were Mexicans and Negroes who
worked on the ranch, but McSpadden rarely gave them dirty
jobs. He also called them by their last names, treated them
with deference, if not respect. He treated them as if they
belonged there. Andrew never felt he belonged anywhere.

In spite of that, he liked the ranch, liked the work, and
he was determined to belong.

Andrew reined in his mount and studied the snowy mud
in front of him. He saw some canine tracks, but ice now filled
them, and it was hard to tell what sort of animal made them.
He truly doubted that there was a wolf anywhere around and
figured that these were probably made by some stray dog or a
coyote. There had been no red wolves seen in this part of
Texas for longer than he had been alive. It seemed incredible
that one had survived the bounty hunters, who for years had
combed the prairies and plains, trapping and shooting them,
bringing in their tails for cash rewards put up by the
Stockman's Association.

On the other hand, Andrew thought when he followed the
indistinct tracks over the crest of a short ridge, if an animal
such as a wolf could survive anywhere, it would be here, in
these redland cedar breaks along the twisting, sandy river bot-

toms. This country belonged to the wolves, he thought. Nothing else could survive there.

The breaks were characterized by a maze of sloughs and impromptu swamps, loblollies and thickets of wild plum and bois d'arc set near clumps of prickly pear. It was so desolate that not even the mesquite invaded it overmuch. Along the narrow, muddy streams that sometimes flowed briny from seeping springs, hardwood trees occasionally found purchase, but they were scraggly and unhealthy looking. No good for farming, the land offered little appeal to anyone except a man of McSpadden's thrifty nature. Along the occasional mini-plateaus and small mesas that rose out of the red loam and its occasional brilliantly white limestone rock, the grass was lush. Every winter the rancher turned his cattle out on it to browse until the winter wheat came up. It made rounding up the heifers and new calves every spring a hard chore, but it saved him money for hay and feed.

Andrew had worked for McSpadden for five years, ever since he left the reservation. Before that, when he finished his mandatory schooling, he looked around and discovered that many of the young men simply stayed there, taking what little work there was, spending their earnings on whiskey and gambling, eventually marrying one of the girls they'd gotten drunk and pregnant, then trying to live on the government dole until the women got fed up and threw them out, went home to their mothers who would greet them with flinging rocks and bricks if they came around seeking a handout. He saw them sitting in front of the reservation store, hunched under cast-off coats and sneaking sips from paperbag-wrapped bottles, trying to blind themselves to their misery. Some were only a few years older than he. Andrew wanted no part of that life, but he had no money, no family and, truly, no future. He needed advice.

He had heard of an old man, Green Glass, who some said was a shaman. He lived out in the mountains and still maintained many of the old ways, although he occupied a tarpaper shack, wore dungarees and workboots and flannel shirts, and

used rubber bands to hold his steel-gray braids tight. It was
said he lived on small game and that he refused to accept anything from the government. Andrew found him. He was half-
Comanche, half-Kiowa, and took one look at this orphaned
boy and called him his son, changed his name from Jackson to
Big Hands. "That was the first thing I noticed about you,"
Green Glass said. "You will come every day and work for me."

None of that was particularly what Andrew had in mind
– especially since Green Glass had no intention of paying
Andrew – but the boy was patient and hoped that eventually
Green Glass would tell him what to do. For three years,
Andrew spent his days working, doing chores around Green
Glass's shack, using the old man's ancient rifle with its splintered stock to hunt rabbits and deer, setting traps for bob-
white, gathering wild onions and asparagus to make stews.
Then he would sit and listen to the old man talk about the old
times, the old ways. Andrew thought a lot of it was boring, and
his questions about what he should do went unanswered. But
he came to like Green Grass, and he felt sorry for him, for his
loneliness and his poverty. The old man, though, never
expressed any regrets about his life. He was fiercely proud,
Andrew learned, and whenever Andrew said anything negative
or uttered a self-pitying remark, then Green Glass's hazel eyes
would harden and his mouth would set. "Such thoughts are
unworthy of you," he said. "Remember who you are."

"Who am I?" Andrew asked, genuinely eager to know.

"You will find out," the old man would reply. Then he
would fall silent and stare into the fire.

When the old man died, Andrew still had no direction for
his life. He felt he had been abandoned again. But he had his
name legally changed to Big Hands in honor of the old man, and
because he had no place else to go, he took up living in the shack.

After a few months, life in the mountains became too
lonely for Andrew to bear. He made up his mind that he was
not meant merely to sit and wait on death to come to him, to
turn into a bitter old man like Green Glass. He wasn't that

much of an Indian, he told himself, and he was young and strong. He had been to school, and he wanted a good job, a white man's job. He had trouble finding one, though, for hard times were everywhere, and unskilled labor wasn't much in demand anywhere near the reservation. It wasn't until he sold everything the old man had left behind and bought a beat-up old truck that he could escape south, across the river and into Texas, where he could find something on his own. It took a while, for no one was interested in hiring an Indian with no trace, but eventually, he signed on with McSpadden's ranch. He now had the security of regular pay for steady, reliable work.

Although McSpadden was good to him, he liked reminding Andrew that he was an Indian. He gave him a horse to ride and a saddle. Even though Andrew had little experience at such arts, he watched the others and learned quickly, and he soon became good with the horses, something McSpadden was fond of pointing out as another part of his natural heritage. Eventually, Andrew took pride in being one of McSpadden's cowboys, even if he was the only Indian among them, and even though they always stood apart from him, sometimes making him feel more lonely than ever.

Today, though, pride was a long way from his mind. He resented being sent out in the frigid weather to search out an animal he only half believed existed simply because McSpadden ordered him to go and wouldn't send anyone else, not even Wilson, who had reported the possibility of a wolf. Any of the other men might have been given the chore, as well, Andrew thought. It might also have waited until this late spring blizzard that was coming on strong had passed. But McSpadden was firm in his decision. The day before, when Wilson came in to tell of his discovery of three steer carcasses down in the breaks, no one had talked of anything else. They were yearlings, he said, and they had been chased down and killed, partly eaten. The only animal large enough to do that was a wolf. Wilson had made the report, but it was up to Andrew to find the wolf – if there was a wolf – and kill it.

Andrew's horse came to the mouth of a deep gulch. The small arroyo cut through the sandy limestone and made its way lazily down in the direction of the river bottom. It probably hit a dead end somewhere before it got there, Andrew thought, noting that the muddy bottom wasn't yet frozen over, meaning that water tended to back out of it in the rainy season. He looked at the ground closely, then dismounted and studied the tracks more closely. They led down around the first turn of the arroyo.

"Maybe it *is* a wolf," he said to his horse, watching his breath vaporize on the frozen atmosphere. "But I don't believe it. Not till I see it."

He climbed back into the saddle and pulled his rifle from its leather boot. It was a nice weapon: a Remington 30.06 with a powerful scope, one of the finest things Andrew had ever held in his hands. McSpadden loaned it to him that morning. The rancher told him he had shot elk, moose, and even caribou with it. Their heads now hung in the ranch house. He told Andrew that if he found the wolf, to kill him and bring him back. He would give him the skin and the tail for the bounty, but the wanted the head for his wall.

It was small enough pay. The breaks were too rough for a pickup or even a tractor in most places. Only a man on foot or on horseback could explore them thoroughly, something the cowboys who worked the spring roundup knew well. Also, Andrew thought, only a man on a horse could hunt down a wolf here. McSpadden reminded him that his Indian blood made him perfect for the task. "You're all natural trackers," the rancher said, slapping Andrew on the back and grinning. "Hell, next to handling horses, tracking and drinking are your best two talents." The other men laughed, and as usual, Andrew grinned back at his employer and nodded. But resentment built like a fever inside him.

Andrew removed the scope's lens caps and looked through it, startled to see how it pulled in distant objects and made them huge in front of his eye. He pulled back the bolt. A shiny brass cartridge gleamed in the breech. Andrew looked again down the twisting arroyo's path, gently spurred his mount forward, and kept the rifle across his pommel, at the ready. A wolf, he figured would run if it could. But the arroyo would narrow. A trapped wolf could be a dangerous thing.

Green Glass had told him many stories about the animals of the prairie. Many, he said, were vital to the People's existence. The old man had never called Indians "Indians." Nor had he called them by their tribal names. He simply referred to them as "People." He said that they had a special connection to the land and to the creatures who lived on it. He told Andrew about times when the land was covered with buffalo, when antelope herds ran so fast and thick that from the height of a mesa, it appeared the very earth had taken life and was moving about. He spoke of when sloughs and breaks such as these were filled with white-tailed deer, beaver, fox, and bear. There were also large and small catamounts, and a variety of hawks, even eagles. The big birds ruled the skies and treetops, Green Glass said. On the earth, the chief of all creatures was the bear, the greatest friend of the People.

But the wolf, he said, was the most highly respected hunter. "The dirt on which a wolf's blood is spilled is sacred," Green Glass told him. "The wolf is the fiercest creature on the earth. It gives no ground, shows no quarter. Always honor a wolf."

But there were no more wolves. Like the bear and the eagle, they were gone. Men had hunted them out, killed them off, and now only cattle and sheep wandered into the breaks, only rabbits and prairie dogs ran from the smaller predators: the coyote, the bobcat, the rattlesnake. "The wolf held out for a long time," the old man said, "but it did no good. They're all gone."

Green Glass shook his head. "The wolf is like the People," he said. "Once we were many, but now we are few. We are poor, and sick, and have no place in the world. The white man came here, hunted us down, starved us and took away our names. They killed us off. Someday, we will all be gone, just as the buffalo and the wolf are gone. Look at you. Look at me. We are People, but we dress and act like white men. We talk like white men. Isn't that strange? But I am old, and I like my comfortable house and soft clothes. I like the white man's whiskey and tobacco, when I can get some, too. I'll bet I'd have liked a lot of his things, even his women. That's a weakness in me. But it doesn't matter. I will soon be gone, too. It is for you I am sad, for you are only a boy. You are an orphan of the earth. You have nowhere to belong. You are like the wolf. You're hanging on, holding out, but someday you, too, will be gone.

Andrew Big Hands guided his mount around the turn in the arroyo, and he wondered why the old man's words came back to him now. He guessed it was because he was now stalking a wolf – or a large dog – and that he resented being ordered to go out in such weather to do it, but he had never taken Green Glass that seriously. He was an old man, but he wasn't old enough to remember the things he spoke of. He was just repeating what he had been told, perhaps by his father. He was bitter and unhappy, Andrew thought. And he wanted Andrew to be bitter and unhappy, too. Then his thoughts softened. Maybe in spite of himself, he was trying to make tracking and killing a wolf a part of his heritage, justifying it somehow by the old man's stories. That was dumb, he thought. I'm an Indian, but that's only an accident of birth. A misfortune, mostly. Green Glass was right: I am only an orphan. None of that means anything compared to work, to money, to a place where I can someday belong.

Overhead the wind and snow picked up even more. When Andrew looked up, he saw a slate gray sky between the rims of the arroyo that now were several feet above his head. He thought he heard something, a long mournful howl. It was only the moaning of the wind in the hackberry and bois d'arc trees, he knew, but for just a moment, he thought it might be the howl of a wolf. It gave him a deeper chill than anything caused by the freezing weather. He gripped the rifle's stock even tighter and patted his horse's neck. He had never seen a wolf except in pictures. His mind told him it was just a dog, a big ugly dog.

In spite of their ruggedness and the difficulty they posed to cowboys each year, Andrew had always liked the breaks, the "badlands," as they were sometimes called by older men. He often rode down into them when he had time to himself. There was a quiet solitude there. The land looked like a Marscape, a red, rugged alien place where time had no meaning. Here, things remained as they were when the People were many and used the land only as they needed it. Sometimes, he believed he could sense their ghosts, their presence all around him. But when he looked at the land with a tougher, modern eye, he had to admit that it was just his fancy that made it seem so. The breaks, wild and remote as they were, were nothing like they had been. Men had come and turned them to whatever advantage they could find. Once, he heard, they had tried to surface mine them for copper, but the ore was of poor quality, so they quit. Then they had drilled for oil in them, and, finding none, left them again. Now, they belonged to those like McSpadden who had stock to run on them. Someday, perhaps, they would find another use for them. But in the meantime, they had to be kept as they were, free of such evil things as wolves. If any still existed.

He urged the horse on, turning now more often as the arroyo deepened and narrowed. He could reach out his arms and touch either side. It grew darker in the shadows beneath the limestone rocks jutting out of the walls. In spite of the coming storm and dropping temperatures, the ground was still warm here, muddy and full of water in places. The canine tracks were more numerous. He was certain that he had found the right trail. The question in his mind was how old the tracks might be, how near a wolf might be, if there was a wolf at all.

Green Glass had told him of how the People hunted. He told him of chasing the buffalo and firing arrows and flinging lances into them, of how brave hunters would leap down from their horses and cut the tendons on the great beasts' legs, causing them to fall. He told of how the hunters would eat the beasts' hearts and livers, drink their blood, still warm, and would smear it on their chests, arms and faces to pay homage to the animal they had slain, of how the women would butcher the meat, take the robes, even collect the bones to be used.

"The People used the land and her creatures well," the old man said. "Today, there is nothing left. Like the People, the whole earth will soon be gone."

Hunting wolves was different, though. When they found a wolf, only one brave would approach it. He would carry only a knife and a lance. He must face the wolf alone. He could bless the animal by his own courage and strength. Then he had the right to wear the skin, to string the teeth into a necklace, or to adorn his chestplate with its claws. Tattoo its symbol on his own skin.

Suddenly Andrew had a vision of himself. He saw himself atop his fine horse, with his powerful rifle and scope ready to fire. All at once, he was disgusted. What would Green Glass say about his heavy coat and wool scarf, or his thick leather gloves and bull hide boots? The old man would be ashamed of him, Andrew thought. He wasn't worthy to kill a wolf. He could give it no blessing of courage. He was no longer a boy, but he wondered at the man he had become. He suddenly decided to stop, to return and explain to McSpadden that the snow was too bad for hunting, the light too poor. Maybe he would send Wilson or another cowboy out tomorrow or next week. Andrew wanted no part of this any more.

Then he saw the wolf.

It was a bitch, and she was huge, the size of two cowdogs, it seemed. She stood with her forepaws over a large hunk of raw meat that looked as if it had come from a calf's hindquarter. She looked up at him when he rounded the last turn of the arroyo and bared her bloody fangs. His horse shied, and he had to grasp the reins tight. He could not turn in the narrow gully, and only with difficulty could Andrew keep the animal from rearing, bucking him off in his fright. He finally steadied, his nostrils flaring, his eyes wide, his breath snorting while his hooves pawed the ground.

The she-wolf growled. Behind her were four pups with coats as red and full as hers. She was gaunt and narrow in her face, but her fur was thick. The whelps also saw and scented Andrew, and joined the fierce show with tiny teeth and small growls.

Wilson was right. There was a wolf. There were whelps and that meant a male was around someplace. And there would soon be more, for the pups looked healthy and would soon grow. He gripped his horse with his thighs, looked at her and set his jaw to keep his teeth from chattering, but he couldn't tell if it was fear or cold that made him shiver.

He had surprised her, and she wasn't sure what to do. He was too far away for her to bound at him and attack, leaving her young exposed to some unseen danger. But he was close enough that she was afraid, angry, and protective. She took a half step backwards, but her fangs still bared. Off to one side, he saw a small opening under a limestone rock: her den, her lair. It was no wonder she had eluded sight before. This arroyo was long, narrow, and given to flooding down near its mouth. But here, the ground was slightly higher, and she could easily stay here, out of sight, venturing out only to eat. To kill McSpadden's cattle.

It was too close for him to use the scope, so he merely pointed the weapon at her. She continued to hold her ground. She twisted her haunches around to protect her pups, yapping and jumping up behind her, never taking yellow eyes off of him.

He sighed heavily. He knew what he must do. He shouldered the rifle and aimed it toward her. He would kill her – the pups as well – then sling her body over his saddle and carry her back to McSpadden's house. The rancher would be pleased. He would give Andrew tomorrow off, maybe a few dollars' bonus to go with the bounty. Maybe he could go into town and buy himself a fine new rifle like the one McSpadden loaned him. Or maybe he could put the money away, save up to buy a new pickup. His old one was so battered and broken down that it was a wonder it ran at all.

All these thoughts ran through Andrew's mind as he looked over the scope and down the barrel of the rifle at the wolf. She was gathering herself into a ball, ready to spring if his horse came forward a step. Her teeth were still bared, bloody and sharp.

He looked down into her eyes. They were the color of prairie grass in autumn, almost the same, he remembered, as the hue of old Green Glass's eyes. Her ears lay back against her skull, and a ridge of red hair rose on her back. He saw her breath vaporizing in the air, almost felt her heat. He believed he could smell her.

"Go on," he whispered to himself, "shoot her and be done with it."

But again he stopped. He looked again into her face, and it was almost as if he saw an expression there. It was foolish, he told himself, for she was only a wolf, a wild animal. But he sensed that she felt something besides fear and anger. It was outrage. To her, he realized, he was an intruder, someone who did not belong here, in her arroyo, in her breaks, near her home. She was angry with him for coming here. This was her place.

Andrew continued to stare at her down the rifle's barrel. She was the last one, he thought. The last red she-wolf in these breaks, in this whole part of the country. She had no beaver or muskrats to hunt any more, no deer or antelope herds to follow, not even buffalo to stalk. All she had was McSpadden's cattle.

Andrew thought of the cattle. They, too, were intruders in her breaks. Big, dumb, ugly, they had plenty of open range up on the prairie pastures, up the fields where McSpadden grew hay and oats and wheat. He invaded the wolf's territory with them, and she had every right to take what she needed to feed her pups. The cattle had no right to be here. She did.

"This is crazy," Andrew said to himself, watching his

words turn to frost when they escaped the wrappings around
his mouth. If he didn't kill her, and if McSpadden found out,
if more cattle were killed, he would be angry. He would fire
Andrew, send him back to the reservation in his broken-down
pickup. He would wind up like Green Glass, poor and living in
a shack on corn mush and coffee he would receive from the
government store. Or worse, he would become like many of his
old classmates, drinking too much and wishing he could die to
ease his loneliness and pain.

All he had to do was pull the trigger. Then go back and
collect his praise from his boss. Show him that even an Indian
can be relied on to do a good job, to return with the goods. It
was simple. It was sensible. It was what he was paid to do. It
was the price of belonging.

Andrew sat still and felt the cold seeping into his skin
beneath his clothes. His horse was growing more and more agi-
tated. He wanted to back away. His head swung this way and
that. Andrew needed to act. But the wolf continued to stare at
him, to growl and snarl. She showed nothing but courage.

If he killed her, he thought, he too would be gone. All the
wolves would be gone. The male, wherever he is, would soon
grow old and die – or he, too, would be hunted down and killed
– and that would be that. She would be like the People, like
Green Glass, like him, Andrew thought. We will all be gone.

At last he lowered his rifle. The wolf stepped forward a
pace, advancing her ground. The horse was backing away
now, and Andrew allowed him to do it. Finally, when the ditch
widened a bit he pulled the reins, turned his mount and gal-
loped out.

"I couldn't do it," he said to the horse when they
emerged from the narrow valley and crossed the ridge, stop-
ping to get his bearings, let the horse blow. "We owe her something better than to die like that." In a way, he thought, he was
practicing what he would say to McSpadden. He knew his boss
wouldn't understand, that he would be angry. Tomorrow, he
would send someone else to kill her, likely. He might come him-

self. He would want the head for his wall.

"But at least it won't be me who kills her," Andrew said. It was more of a promise than anything, he thought. Then he realized that he had found the wolf only by luck. If he didn't tell where she was, there was a good chance that no one would find her when the snow melted and her tracks faded away into the mud.

He glanced back into the arroyo one more time. Its twists and turns disappeared into the gathering darkness. Snow was now falling heavily, covering the tracks. By next spring, her pups would be grown, maybe have litters of their own. Maybe it would be a long time before anyone found any of them. Maybe by then the breaks would be filled with wolves, and McSpadden would keep his cattle out of there. Maybe the deer would come back, too, and even the antelope. It was too late for the buffalo and the bear, but if the wolf was there, maybe other creatures were, too. Maybe, he thought, the earth could heal itself, if it was left alone, if man could be kept out and away, maybe something of its honor could be saved. It was a faint hope, but it was all there was.

Andrew turned his horse back toward the ranch house and lowered his head against the weather. Snow now covered him completely as he climbed out of the breaks and into the pastures bordering it. He spied lights in the distance where the men played dominos and laughed in the warmth of the bunkhouse. He didn't think about what he had done – had not done. He only thought about the People. Somehow, he thought that so long as the wolf lived, the People lived. It made no sense – not logically – but in a way, he was sure that Green Glass would agree.

As the lights of the ranch grew brighter through the snowy gloom, he had another thought. He decided to change his name to Andrew Many Wolves. He hoped that it would be a good omen, and he, once again, felt sure that Green Glass would agree.

Afterword

Rick Bass

This afternoon a **golden eagle** wheeled past my upstairs window in a swirling snowstorm, playing in the fifty-mile-an-hour gusts that bend the big trees and snap the tops off smaller ones. It got me to thinking about where I grew up, in Houston, on Buffalo Bayou. Where I live now, in the Yaak Valley, the most northwesterly valley in Montana, I can walk due north through the snow for a few miles and cross over into Canada. The Yaak is one of the wildest valleys in the Lower Forty-Eight, a place so wild that we've still got fifteen or so grizzly bears living here, one of only six places in the United States where that supremely threatened species is still found. I've seen mountain lions on my driveway, elk and moose in the marsh, grizzly and wolf tracks in the front yard. It is a place of extreme biological diversity and splendor, Montana's only rainforest. I realize that it was growing up alongside Buffalo Bayou, next to Houston's Memorial Park nearly half a century ago now, that prepared a place in my heart for the Yaak.

When that eagle wheeled past, I had been looking at the newly proposed master conservation plan for Memorial Park. Memorial Park may not quite possess the old-world elegance of New York's Central Park, but it is in many ways more vital and vibrant, in the rank and unruly way of bayous, with their subtly-stirring pulse of life, the slow-moving current laden with biological riches as the bayou draws nearer to the sea and the end of its heroic natural journey.

In reviewing the master plan, I saw mention of where Chevron had tacked on to the park, donating seven acres.

Laudable gifts, the worth of which I'm sure tallies well into the
millions, or, for all I know, billions, these days. (In some parts
of Montana, those seven acres might still be gotten for a cool
thousand dollars). I consider myself lucky to live where I do
now – forty-below winters notwithstanding – and reading from
a distance of 1,500 miles about Houstonians' efforts to beef up
and protect the slender ribbon of green that bisects their city
– our city – I realize how lucky I was to have had that sinuous
park while growing up. If I had spent my hours and days collecting spent bullet casings from the gutters of inner city
streets, collecting them to make into jewelry, perhaps, might I
have grown a harder heart than that of the boy who slipped
down to the bayou whenever he could and peered through the
ferns and yaupon, mesmerized by the sun-dappled ghosts of
alligator gars and giant soft-shelled turtles floating on the surface of that brown water, and by the wild violets that grew
amongst the pine needles and oak leaves? Even today the
black-light cobalt intensity – the ultraviolet of those violets –
feels etched onto my retina, and though that section of woods
is long gone, the peace it emblazoned on my mind is not, and
it occurs to me that a good day up here in Montana is one in
which some marvel of a wilder, perhaps grander nature
impresses itself upon me with even a proximate degree of the
same peace that I knew as a child, an innocence and peaceful-
ness that I suspect was so much richer, back then – as dense as
that cobalt color – for the unquestioned nature of it.

These days, of course, there's often a grand arrival,
grand entrance, that attends to the infrequent noticing of such
peace by adults, so harried and fragile and fragmented have
we become. Though even that is all right, more than all right.
We take whatever peace that parks, and open space, and wild
nature – or, some days, any kind of nature at all – can bring
to us, having traveled so far from the territory of childhood.

What uncoiled with me, over the years, so that eventually in order to know a deeper peace, I required the experience
of tens or even hundreds of thousands of acres, where once

only six or seven had been sufficient? Such mysteries are perhaps only of interest to our own individual selves, intensely personal and intensely specialized: but they matter. I find it wonderful that nature in no way means the same thing to any two of its admirers. For my own part, however, it is when I am in the wilderness that I perceive, seemingly all of a sudden, that despite being insignificant in the world, I am most fully engaged with the world. Watching a sunset, or standing in an old forest, I feel, paradoxically, despite my insignificance, more fully a participant: and in that realization, it will occur to me then that for some time, without my even knowing it, I had not been feeling that inclusiveness.

※

I sometimes wonder if this issue of the universal loneliness that whispers to all of us in one form or fashion, and visits each of us at one time or another, has often to do with the fact of whether we are willing to accept ourselves as a part of the earth, or apart from it. It's a big question, and I could be wrong.

In such remembrances of our utter insignificance, and of the natural world's grandeurs, a certain fundamentalist fervor will come over me in a wave, briefly. The words of Ecclesiastes might come to mind, where it is written, "For that which befalleth the sons of men befalleth beasts; even one thing befalleth them: as the one dieth, so dieth the other; yet, they have all one breath; so that a man hath no preeminence above a beast" – or even the jeremiad of Jeremiah himself, who lamented " The peaceable habitations are cut down" – but mercifully, then – or at least I think it is a kindness, a mercy – the moment will pass, and I will be lulled like so many of the rest of us back into thinking that nature is no big deal, and that there is no mystery, no divinity behind the veil: no miracle in the scrimwork of beetles, no divinity in the design of a single feather or a single seed. In my return to that lulled condition, the moun-

tain will appear once more as only a mountain, and the swamp
as only a swamp.

So what if we hacksaw the topskull off of the mountain,
or drain away the swamp's heart? It's just a living thing, and
it's certainly nothing nearly so grand as any one of our own
wonderful bright-burning short-lived individual selves, six bil-
lion of us marching across those mountains, marching across
those swamps and through those forests, yearning for a kind
of eternity that the earth already possesses, has always possessed in the seed-heart of the universe.

⸺⸺⸺

Other times, I snap out of that lull, that narcotized twen-
ty-first century trance, and ask myself, What is holy, and what
is the nature of hope, and how does one conduct one's self in
the presence, the midst, of certain loss?

Against such crush of domesticity – of homogeneity, and
the destiny of more and more concrete – does one seek to main-
tain pragmatic dignity, and to achieve successes where we can,
in scattered little six- and seven-acre gardens? If the wild
beauty of the made and physical natural world is lost or further compromised, then how much longer can the echoes of
that beauty and innocence remain kindled in our own hearts?
I suspect we won't know the answer until it's all gone: which
makes, perhaps, those six-acre parks, and those 60,000-acre
wildernesses, and those 6 million-acre ecosystems, all the more
important; and which is what makes the art – the voices and
the minds – of the writers gathered here all the more important, as well.

I think also that beyond the questions or issues of dignity
and self-preservation, there is an issue of reciprocity at play
here. Call it debt, or honor, or noblesse oblige, or what-have-
you, but it seems daily that there is a code of ethics, a code of
conduct, with which we – Texans, Montanans, and all the rest
of us – should address the earth, but that we have moved

steadily away from that allegiance. Have been lured away, or distracted.

Beyond the glorious basics of having experienced the birthright of having known what it's like to breathe clean air, to drink clean water, and to eat food not poisoned by any manner of toxins, I personally am grateful to the natural world for the gifts and lessons of imagination: gifts and lessons I've received in Texas as well as in Montana. Although I am many years removed from my childhood there, I am not so distant from the imagination which the park, and its living pulse, the bayou, nurtured in me, nor my gratitude to the landscape for that nurturing, for it has helped me pursue my avocation or calling of learning to become a writer: a task that begins anew each day, but whose genesis came in large part from having and knowing that green space, that open space, as a young person growing up in Houston during a time of turbulent change.

The wolves, elk and grizzlies that surround my cabin in Montana now are part and parcel, in my mind, and in my life, of the same fabric of life I grew up with near Memorial Park: the clatter of dry oak leaves in the fall during a weather-change; the delicate patter of skinks rustling on the forest floor; the summer-scold of fox and gray squirrels, the flashing tail of a deer, the quietude of rabbits, the ornamentation and endurance of box turtles . . .

I believe that landscape exists for reasons other than the immediate service it may provide to humans – whether for clean air and water filtration, noise abatement, spiritual renewal, scientific discovery, and the provision of solitude and solace. Landscape is in my opinion its own element in the world, deserving of protection and respect beyond the values it provides us both in the short term as well as long term – but when I think of Buffalo Bayou, I cannot help but think self-ishly of all that it has given me, at all the different times of my life, in all seasons and emotions, and I like to think that there are many others, in far different professions and walks of life, who have also benefited from the bayou's grace and presence.

I like to think that the sum, the tapestry, of those benefits is quite a remarkable and durable, wondrous thing, like that of a great city, culture or civilization.

Those of us who were fortunate enough to be children in Texas were "wild and powerful in our innocence," as Paul Christensen writes here in his eloquent introduction. I remember passing by "Wolf Corner," out on the Katy Prairie, each Sunday morning on the way to church, and viewing that week's tally of coyotes and red wolves which had been trapped and shot and poisoned, then flung over the barbed-wire corner fence post: a rogue's gallery of wildness hanging upside down as if watching us on our way to church, placed there perhaps as illuminated example of how near lay the errant path of sin – how fiercely narrow that path was, and how wild and ravening the rest of the world beyond still was, to the west, at least, if no longer in all directions.

The wind would ruffle the fur of those dead beautiful creatures. They were almost always there – the land seemed to keep giving up an inexhaustible supply of them – until finally, as I grew older, it was not the wolves and coyotes that disappeared, but the prairie at Wolf Corner itself: though still I did not know rage, nor really even sadness. I knew only hope and imagination, believed only in boundlessness and possibility – and always, it was the image, the fabric, of a wild and uncompromising nature – or my perception thereof – that nurtured such seeds in me.

I was not always in a wilder nature, while growing up, but I saw it, frequently enough, and knew – believed – it was out there. That other world emanated from the lonely wind-stirred tails and neck-ruffs fluttering there on the fence at Wolf Corner, and from the soothing images of seemingly-endless rolls of the unbroken oak-and-juniper hills that framed the Balcones Escarpment, hills like mythic castle-walls rising above the jeweled city of Austin: a wilder, more heavenly nation lying to the north and west of there, and which barbarians could surely never storm. Here, too, as with Wolf Corner, I was mistaken – there was no permanence – but what was important, I submit, and valuable, was not "just" the gift of clean air and water and solace which that ill-fated open space bequeathed to me and every other Texan, but the gift of imagination, too: the simple, priceless lesson – a lesson that cannot be replicated – that Here is a place still unlike other places.

It was a lesson in enthusiasm – an enthusiasm for diversity – that was taught to all children, and all adults, growing up within sight and reach and knowledge of a swamp, or a redrock desert, or a Caprock vista, or a high plains prairie; an oak savannah, a juniper mesa, a Gulf tidal flat, a foggy pinewoods red-clay forest, a Big Bend without air pollution. It seemed to me that there was more of Texas, in those days, and that now there is less.

The mere act of witnessing involved an embracing of the virtues of gratitude. The wilder land helped make us into better people – it asked the best of us – though being human, we sometimes gave it our worst; sometimes intentionally, other times unintentionally. We have let too much of it – nearly the last of it – slip through our fingers.

There is still much to love, as many of the writers in this volume testify. There will always be much to love. And in a time of war, are not those of us who hail from this most warrent of states more in need than ever of finding issues and topics upon which to exercise that most human of talents, love? No other state in the Union possesses the blood history, the heritage of violence and disrepair, from which present-day Texas was born: no other state – not even the modern nation of Israel, as Texas historian T.R. Fehrenbach has pointed out, fought three running wars at once, during the genesis of statehood: the Indian wars, the Civil War, the War against Mexico, with aggressive sorties against us by Great Britain and Spain thrown into the mix for good measure.

If nothing else, nature teaches us that where we come from is what and who we are, and from such a history of tear-ing-apart, it seems to me to be past time, these 168 years later, to roll up our sleeves and begin considering new stories of reassembly. Any culture can tear down or otherwise ransack the treasures that preceded that culture's arrival. The real talent lies in putting broken things back together. And in the natural world, in Texas, we have an ample list of broken-apart things to choose from as we consider – if we consider – the reassembly.

Each of the writers in this volume has, I suspect, his or her favorite landscape, his or her most clamant issue. With regard to the questions of "What to save?" and "Where to act? Where to begin the reassembly?" – I suspect that perhaps the most compelling case to be made might involve that most necessary and precious and limiting of resources, water – for the most part, that which we have not pumped dry and squandered, we have poisoned or at least tainted – but that is a concern of the flesh, and while we still have the brief opportunity to consider such indulgences, I would like to make a case for preserving not just the concerns of our flesh – clean air, and clean water – for those challenges will be coming with ever-increasing frequency, soon enough – but of the spirit, as well, and of the need, the legacy as well as birthright, of Texans to have open space: for our minds, our imagination and our sanity, in addition to serving as an anchor and stabilizer upon all our other needs.

This, as much as the native wildness of the West, is what I am most grateful for. In wild nature and the open spaces of the West, I still experience the pure democracy of nature, with the full sophistry of billions of years of accords and discords integrated into one moment, and one place; and, given sufficient space, I sometimes feel a part of that natural, more graceful and yet mysterious world.

It remains an elegant world, and for me, at least, and many others, it remains a portal, an avenue, to the world of imagination. I've often wondered at this paradox, at how something so primal and elemental, so tangible and physical, can be linked so inextricably to its seeming shadow-opposite, the ethereal of the imagination. It seems to make no sense, to defy logic: how the sight, the fact, of the bioluminescence of a firefly, or the touch of a curve of granite, or a certain clean scent of beach salt – the tangible, in all its essential glory – can ignite the intangible of the imagination, both in memories and in dreams – but time and again it does, as if one must feed the other. And whether the physical world depends upon the spirit, or the spirit upon the physical, the world's philosophers have wrangled forever, with no clear decision. For my own part, I would choose to hedge my bets, and proceed on the assumption that they feed each other.

How, over such a relatively short period of time, did we Texans go from being one of the most wide-open, unbounded and creative states to one of the most congested and space-impoverished, with the least amounts of open-space per capita among our urban centers of any state in the country? How have we allowed ourselves to run nearly out of water, how did we convert freedom and strength into captivity, dependence, vulnerability and weakness?

And just as importantly, what path might exist, or be made, to lead us out of this current situation, and back into a culture of strength and possibility, if not innocence?

Fehrenbach has noted that historically, Texans viewed land rather than knowledge as the source of all wealth. In the old days, when there was so much bounty – the spoils of war – that seems certainly to have been the case. And more recent economic models seem to suggest that the latter model is the one that bears the most fruit now.

But I want to suggest that we consider a third way, a sort of land-spirit hybrid – a more mature and experienced melding of the two – and to suggest that a culture that makes a new commitment to open spaces, and the creative spirit, and acknowledges with greater humility – not always a typical Texas virtue, I understand – the mysteries and power and, indeed, sacredness of nature might in the long run become the

wealthiest yet – economically, perhaps, but spiritually, physi-
cally and emotionally, almost for certain.

These kinds of discussions – these yearnings – cannot be
jump-started into the sausage-grind of legislation; cannot be
fitted, at first, into code and structure. In their nascent
reawakening, they have little – for now – to do with tax law or
subsidies. The yearning is beneath all these things and is somehow a part of these things, but where it begins, or re-begins, is
in the spirit, and in in conversations among friends and family and neighbors. The old Texas virtue of storytelling and
imagination has perhaps its greatest challenge – how to put the
pieces back together, how to fix our damaged things? In our
entire vast and great state, we have only one remaining unpro-
tected roadless area of public land greater than 4000 acres in
size. In this, we are one of the poorest states in the nation.
How did we go from being the richest to among the poorest,
and what does it portend for our future, if we do not address
this absence as well as this hunger?

In Montana, we are still arguing about specifics – about
the fate and importance of a certain species of trout, a certain
and magnificent species of bear, a certain and magnificent
type of forest.

In Texas, I think we need to start almost all the way over,
however, and renew a dialogue with the basic elements. Air,
water, earth, space, and the spirit, the fire, that emanates
from the conjunction, the divine weaving, of these things, and
which has made each of us who we are.

My gratitude goes to the editor of this book, Paul
Christensen, who assembled and edited the abstract idea of it,
and to the publisher, Bryce Milligan of Wings Press, and particularly to all the contributors, who continue to integrate
these elements into the weave of their stories, and into the passions of their work, their lives, their communities, their government. We can never go back to where we were, but we can
dream and build new cultures of generosity, tolerance and
vision. We can still, even at this late date, transcend nostalgia,

loss and regret, can transcend even the burden of longing and make our way like long-lost travelers back to a real and physical world of vermilion flycatcher, black bear and bunting.

It is a long way back to such freedoms, this path from the imagined back to the real, but it is a journey the writers in this book believe is worth traveling, and have begun; and even in these often dark and frequently frightening days I am comforted when I consider how easy and natural the first step of such a journey is, and where in each of us it begins. There is the grand mystery of outer space, beyond the arc of the atmosphere, and, at present, beyond the arc of knowing. But there is the mystery too – and the gift of imagination – in the more modest arc of space that extends, still, in some places, to something as simple yet rare as an uncontested, undeveloped or undamaged horizon, or a grove of green.

There are things still to be found in that open space, or in that grove, or along that creek – valuable things, deserving of honor and respect – and seen from a distance, it pleases me to realize that Texans are still laboring, sometimes against long odds, on their behalf.

Contributors

Michael Adams teaches fiction writing at the University of Texas at Austin. Among his books are *Blindman's Bluff* (1982) and *Anniversaries of the Blood* (1988).

Wendy Barker is a poet and critic, and teaches literature at the University of Texas at San Antonio. Among her books of poems are *Winter Chickens* (1990), *Let the Ice Speak* (1991), and *Way of Whiteness* (2000), winner of a Violet Crown Award from the Writers' League of Texas. Critical works include *Lunacy of Light: Emily Dickinson and the Experience of Metaphor* (1987) and *This House Is Made of Poetry (1996)*, co-edited with Sandra M. Gilbert, on the poetry of Ruth Stone.

Rick Bass is a widely published essayist and novelist on environmental subjects, particularly the Yaak Valley of Montana where he lives. Recent books include *The Book of Yaak* (1996), *The New Wolves: The Return of the Mexican Wolf to the American Southwest* (1998), *and The Roadless Yaak: Reflections and Observations about One of Our Last Wilderness Areas* (2002). In 2005, Houghton Mifflin will publish his new novel.

Michael Berryhill is a journalist who writes about prison culture and social justice.

Paul F. Boller taught history at Texas Christian University and is the author of many books on American politics, including *Presidential Anecdotes* (1996) and *Presidential Campaigns* (2004).

Jerry Bradley is Dean of Graduate Studies at Lamar University in Beaumont, Texas. Among his books are a collection of poems, *Simple Versions of Disaster* (1991) and *The Movement: British Poetry in the 1950s* (1993).

Mark Busby, past president of the Texas Institute of Letters, is director of the Center for the Study of the Southwest at Texas State University at San Marcos. Among his books are *Fort Benning Blues* (2001), a novel; *Ralph Ellison* (1991), and *Larry McMurtry and the*

West: An Ambivalent Relationship (1995). He is the editor of *From Texas to the World and Back: Essays on the Journeys of Katherine Anne Porter* (2001), among many others works.

Pat Carr taught at Rice University and the University of New Orleans. Among her books are the novel, *Night of the Luminarias* (1986), and *Sonahchi: A Collection of Myth Tales* (1988). She lives in Elkins, Arkansas.

Rosemary Catacalos directs Gemini, Inc., in San Antonio, which brings national and regional writers to Texas. She was a Stegner fellow in poetry at Stanford, and her books of poetry include *As Long as It Takes* (1984) and *Again for the First Time* (1984).

Paul Christensen is a poet and writer, and teaches modern literature at Texas A&M University. He shared a "Distinguished Prose" award with Rick Demarinis from *Antioch Review;* received the "Violet Crown" award from the Writers' League of Texas for *Blue Alleys: Prose Poems* (2001); and was an NEA Poetry fellow in 1991. His most recent books are *West of the American Dream: An Encounter with Texas* (2001) and a book of poems, *The Mottled Air* (2003).

Betsy Colquitt is a retired professor from Texas Christian University, and the editor of the quarterly journal, *Descant.* Among her books of poetry are *Honor Card and Other Poems* (1980) and *Eve–from the Autobiography* (1997).

Bill Crider is a novelist whose books include *A Romantic Way to Die* (2002). He lives in Austin.

Elizabeth Crook is a novelist and author of *Promised Land* (1994) and *The Raven's Bridge* (1996). She lives in Austin.

Tracy Daugherty is a fiction writer and professor of English at Oregon State University, at Corvallis. His work includes *What Falls Away* (1996) and *The Boy Orator* (1999), and the short story collection, *The Woman in the Oilfield* (1996).

Robin Doughty teaches in the Geography Department at the University of Texas at Austin. Among his books are *The Eucalyptus*, *Return of the Whooping Crane*, and *Endangered Wildlife in Texas.*

Laura Furman writes fiction and essays, and teaches writing at the University of Texas at Austin. Among her novels are *Tuxedo Park* (1983), *The Glass House* (1986), and the memoir, *Ordinary Paradise* (1998). She founded the journal, *American Short Fiction*, and edited the *O'Henry Prize Stories* in 2003.

Reginald Gibbons has published seven books of poetry, including *Sparrow: New and Selected Poems* (1997), *Homage to Longshot O'Leary* (1999), and *It's Time* (2002), winner of the Natalie Ornish Poetry Award from the Texas Institute of Letters. His fiction includes short story collections, notably *Five Pears or Peaches* (1991) and the novel, *Sweetbitter* (1996). He teaches literature and writing at Northwestern University.

A.C. Greene (1932-2002) was a noted journalist and historian, and an influential critic of Texas literature. He wrote for the *Dallas Morning News* and the *PBS* show, *Newshour,* and served as the Coordinating Director of the Center for Texas Studies at the University of North Texas in Denton. Among his books are the memoir, *A Personal Country* (1969), *The Fifty Best Books of Texas* (1981), and *Taking Heart* (1990).

Pete A.Y. Gunter writes on ecology and teaches philosophy at the University of North Texas. His books include *The Big Thicket: A Challenge for Conservation* (1972) and *The Big Thicket: An Ecological Reevaluation* (1993). He co-authored *Texas Land Ethics* (1997) with Max Oelschlaeger.

Stephen Harrigan is a novelist and screen writer, and a former senior editor with *Texas Monthly.* Among his novels are *Aransas* (1980), *Jacob's Well* 1984), and most recently, *The Gates of the Alamo* (2000). He has written screenplays for HBO and CBS, and teaches writing at the University of Texas at Austin.

William Harrison is a novelist whose books include *In a Wild Latitude* (1969) and *The Blood Latitudes* (1999).

William Thornton Hauptman is a playwright and novelist. Among his dramatic works are *Domino Courts/Comanche Cafe* (1978), *Big River* (1985), a musical based on *The Adventures of Huckleberry Finn*, winner of Tony Award for Best Musical Book. His short story collection, *Good Rockin' Tonight and Other Stories* (1986) won the Jesse Jones Award for Best Fiction given by the Texas Institute of Letters. His novel, *The Storm Season*, appeared in 1992. He teaches writing at the Texas Center for Writers at the University of Texas at Austin.

Olive Hershey is a novelist and short story writer. She is the author of *Truck Dance*; her stories have appeared in numerous journals and reviews.

Steven G. Kellman teaches comparative literature at the University of Texas at San Antonio. Among his books are *The Self-Begetting Novel* (1980) and *The Translingual Imagination* (2000).

Pat Little Dog has been a bookstore owner and a small press publisher. A short story writer, essayist, and poet, her books include *Border Healing Woman: The Story of Jewel Babb* (1981), and a collection of stories, *Afoot in a Field of Men* (1988).

Bryce Milligan took the cover photograph and designed *Falling from Grace in Texas*, which is the 50th book published by his Wings Press. An award-winning poet, song writer, children's author, young adult novelist, and anthologist, he is the author of a dozen books. He has also edited several important collections of work by Latina authors, including *Daughters of the Fifth Sun* (1995).

Charlotte Baker Montomery received the Cokesbury Award from the Texas Institute of Letters for her books, *Magic for Mary M.* (1953) and *Best of Friends* (1972). She has written and illustrated many children's books and is active in the Humane Society.

Michelle Pulich is an environmental activist and an outreach coordinator with Living Rivers, an ecological group in Arizona concerned with the health of the Colorado River.

Clay Reynolds is a novelist and short story writer, who teaches writing at the University of Texas at Dallas, where he is Associate Dean of undergraduate studies. Among his novels are *The Vigil* (1986, 2001), *Franklin's Crossing* (1992), and *Players* (1992), *Monuments* (2000), *The Tentmaker* (2002). His latest novel is *The Whore of Hoolian*, due out this year.

Pattiann Rogers' newest book of poems is *Generations*, due out this year. Among her ten books are *Firekeeper, New and Selected Poems* (1991), winner of the Natalie Ornish Poetry Award from the Texas Institute of Letters. Her earlier books are collected in *Song of the World Becoming: New and Collected Poems, 1981-2001*. She lives in Colorado.

Lisa Sandlin is a novelist, and author of *The Famous Thing About Death* (1991) and *Message to the Nurse of Dreams* (1997), winner of the Violet Crown Award for literature from the Writers' League of Texas and the Jesse Jones Award for fiction from the Texas Institute of Letters. Her newest book is *In the River Province* (2004). She teaches literature at Wayne State University.

Bud Shrake is a novelist and sports journalist who has written for the *Dallas Morning News* and *Sports Illustrated*. Among his seven novels are *Blessed McGill* (1968), *Night Never Falls* (1987), and *Billy Boy* (2002).

Carmen Tafolla is one of the most often anthologized Latina writers in the country and a highly sought-after speaker and performer. Her most recent books are *Sonnets and Salsa* (2001) and *Baby Coyote and the Old Woman / El coyotito y la viejita* (2002). She is the founding director of the Camino School for the Gifted and Talented in San Antonio.

Cathy Tensing teaches reading and analysis in the English Department at Texas A&M University.

Marshall Terry is a retired Distinguished Professor of English at Southern Methodist University, Dallas, and a fiction writer. Among his novels are *Ringer* (1987), *My Father's Hands* (1992), and the collection, *Dallas Stories* (1987). In 1991 he received the Barbara McCombs/Lon Tinkle Award for "excellence in letters" from the Texas Institute of Letters. He also won the Jesse Jones and "Best Short Fiction" awards from the TIL.

Ron Tyler teaches history at the University of Texas at Austin, and directs the Texas State Historical Association. He is editor of *The New Handbook of Texas* (1996), and the *Southwestern Historical Quarterly*.

Charlotte Whaley, a historian, is the author of a biography, *Nina Otero-Warren of Santa Fe* (1994) and the novel *Mybia Cartier* (1991).

Bryan Woolley has long been a feature writer with the Dallas Morning News, and is a prolific author on Texas subjects. Among his recent books are *Sam Bass* (2004), *Generations and Other Texas Stories* (1998), and *Home is Where the Heart is*.

Michael Zagst is a novelist and screen writer, whose work includes *The Greening of Thurmond Leaner* (1986), *The Sanity Matinee* (1987), and *The Last Man in Paradise* (1994).

Acknowledgments

The following pieces were published previously, and the editors wish to express their thanks to the publishers and editors for permission to reprint them here:

Mark Busby: "The Sacred Hoop" first appeared in *New Texas 2000* (Denton: University of North Texas Press, 2000).

Jerry Bradley: "How the Big Thicket Got Smaller," appeared in *Simple Versions of Disaster* (Denton: University of North Texas, 1991).

Rosemary Catacalos: "Homesteaders" first appeared in *Again for the First Time* (Tooth of Time, 1984) and has been reprinted many times since.

Paul Christensen: "Water," first appeared in *Madison Review* (1995).

Tracy Daugherty: "Amarillo," appeared in *Five Shades of Shadow* (Lincoln: University of Nebraska Press, 2004).

Laura Furman: "The Woods," appeared in *Drinking with the Cook* (Houston: Winedale Publishing, 2001).

Reginald Gibbons: "Refuge" and "Manifesto Discovered under a Fringed Gentian" appeared in *It's Time* (Baton Rouge: Louisiana State University Press, 2002).

Stephen Harrigan: "The Seam of Water and Land," first appeared in *Audubon Magazine* (July-August 1993).

Clay Reynolds: An earlier version of "The Last Wolf" appeared in *Potpourri* (1997).

Pattiann Rogers: "Being Accomplished" and "A Passing" first appeared in *Song of the World Becoming: New and Collected Poems, 1981-2001* (Milkweed Editions, 2001).

Marshall Terry: "Lead (Pb)," first appeared in *New Texas '91* (Denton: University of North Texas, 1991).

Ron Tyler: "Smog on the Horizon," appeared as the introduction to *Pecos to Rio Grande: Interpre-tations of Far West Texas by Eighteen Artists (Texas A&M University Press)*.

Colophon

This first edition of *Falling from Grace in Texas: A Literary Response to the Demise of Paradise*, edited by Rick Bass and Paul Christensen, has been printed on 70 pound paper containing fifty percent recycled fiber. Text and titles have been set in a contemporary version of Classic Bodoni. The font was originally designed by 18th century Italian punchcutter and typographer, Giambattista Bodoni, press director for the Duke of Parma. Initial capitals were set in Caslon Openface Type. All Wings Press books are designed and produced by Bryce Milligan. Elements of the text organization in this book were suggested by Robert Bonazzi.

Other recent titles
from Wings Press

The Angel of Memory by Marjorie Agosín

Way of Whiteness by Wendy Barker

Hook & Bloodline by Chip Dameron

Incognito: Journey of a Secret Jew by María Espinosa

Peace in the Corazón by Victoria García-Zapata

Street of the Seven Angels by John Howard Griffin

Black Like Me by John Howard Griffin

Cande, te estoy llamando by Celeste Guzmán

Winter Poems from Eagle Pond by Donald Hall

Initiations in the Abyss by Jim Harter

Strong Box Heart by Sheila Sánchez Hatch

Patterns of Illusion by James Hoggard

With the Eyes of a Raptor by E. A. Mares

This Side of Skin by Deborah Paredez

Fishlight: A Dream of Childhood by Cecile Pineda

The Love Queen of the Amazon by Cecile Pineda

Bardo99 by Cecile Pineda

Face by Cecile Pineda

Smolt by Nicole Pollentier

Prayer Flag by Sudeep Sen

Distracted Geographies by Sudeep Sen

Garabato Poems by Virgil Suárez

Sonnets to Human Beings by Carmen Tafolla

Sonnets and Salsa by Carmen Tafolla

The Laughter of Doves by Frances Marie Treviño

Finding Peaches in the Desert by Pam Uschuk

One-Legged Dancer by Pam Uschuk

Vida by Alma Luz Villanueva

Wings Press Anthologies:

Cantos al Sexto Sol: A Collection of Aztlanahuac Writing
edited by Cecilio García-Camarillo, Roberto Rodríguez,
and Patrisia Gonzales

Jump-Start PlayWorks
edited by Sterling Houston

Falling from Grace in Texas:
A Literary Response to the Demise of Paradise
edited by Rick Bass and Paul Christensen

2005-2006 titles from Wings Press

Among the Angels of Memory / Entre los ángeles de la memoria
by Marjorie Agosín

The Scribbling Cure by Robert Bonazzi

Psst . . . I Have Something to Tell You, Mi Amor
by Ana Castillo

Drive by Lorna Dee Cervantes

Tropical Green by Chip Dameron

Tracking the Morning by Robert Fink

Burying the Farm by Robert Flynn

Wearing the River by James Hoggard

Inventing Emily: The True History of the Yellow Rose of Texas
by Denise McVea

Freize by Cecile Pineda

Indio Trails by Raúl Salinas

Songs Older Than Any Known Singer by John Phillip Santos

Scattered Risks by Pamela Uschuk

Our complete catalogue is available at
www.wingspress.com